The Chemistry of the Atmosphere: Its Impact on Global Change

Perspectives and Recommendations

John W. Birks, Editor
University of Colorado

Jack G. Calvert, Editor
National Center for Atmospheric Research

Robert E. Sievers, Editor
University of Colorado

Plenary Lectures and Recommendations from the
International Conference on the Chemistry of the Atmosphere
Sponsored by the International Union of Pure and
Applied Chemistry and the American Chemical Society,
Baltimore, Maryland, December 2–6, 1991

American Chemical Society, Washington, DC 1993

brary of Congress Cataloging-in-Publication Data

ernational Conference on the Chemistry of the Atmosphere (1991: Baltimore, Md.)   The chemistry of the atmosphere: its impact on global ange: perspectives and recommendations / John W. Birks, Jack G. Calvert, Robert E. Sievers, editors.

p.   cm.

"Plenary lectures and recommendations from the International Conference on the Chemistry of the Atmosphere, Baltimore, Maryland, December 2–6, 1991."

Includes bibliographical references and index.

ISBN 0–8412–2532–X.—ISBN 0–8412–2533–8 (pbk.)

1. Atmospheric chemistry—Environmental aspects—Congresses. 2. Climatic changes—Environmental aspects—Congresses.

I. Birks, John W. II. Calvert, Jack G. (Jack George), 1923–   . III. Sievers, Robert E. IV. American Chemical Society. V. Title.

QC879.6.I55 1991
551.5′11—dc20   92–41499
                                                                              CIP

The paper used in this publication meets the minimum requirements of American National Standard for Information Sciences—Permanence of Paper for Printed Library Materials, ANSI Z39.48–1984. ∞

PRINTED IN THE UNITED STATES OF AMERICA

Second printing 1993

CONTENTS

ACRONYMS

ASCEND-21	Agenda of Science for Environment and Development in the 21st Century
BAPMoN	Background Air Pollution Monitoring Network (WMO)
CFC	Chlorofluorocarbon
CHEMRAWN	*CHEM*ical *R*esearch *A*pplied to *W*orld *N*eeds
CMDL	Climate Monitoring and Diagnostics Laboratory
EMEP	Long-Range Transmission of Air Pollutants in Europe
EPA	(United States) Environmental Protection Agency
EPRI	Electric Power Research Institute
FCCSET	Federal Coordinating Council for Science, Engineering, and Technology
GAW	Global Atmospheric Watch (WMO)
GCM	General Circulation Model
GEMS	Global Environment Monitoring System (UNEP)
GHG	Greenhouse Gas
GTC	Gigatons Carbon
GWP	Global Warming Potential
HCFCs	Hydrogen-containing Chlorofluorocarbons
HFCs	Hydrogen-containing Fluorocarbons

HTR	High Temperature Reactor
HVDC	High Voltage Direct Current
ICSU	International Council of Scientific Unions
IGAC	International Global Atmospheric Chemistry Program
IGBP	International Geosphere-Biosphere Program
IMAP	International Middle Atmosphere Program
INC	Intergovernmental Negotiating Committee
IPCC	Intergovernmental Panel on Climate Change
IUPAC	International Union of Pure and Applied Chemistry
MAP	Middle Atmosphere Program
MNCs	Multinational Corporations
MST	Mesosphere-Stratosphere-Troposphere
NAPAP	National Acid Precipitation Assessment Program
NAS	National Academy of Sciences
NASA	National Aeronautics and Space Administration
NOAA	National Oceanic and Atmospheric Administration
RLEM	Research Laboratories for Materials and Structures
SCOPE	Scientific Committee on Problems in the Environment
START	System for Analysis, Research and Training
STIB	Stratosphere-Troposphere Interactions and Biosphere (Program)
UARS	Upper Atmosphere Research Satellite
UN	United Nations
UNCED	United Nations Conference on the Environment and Development
UNEP	United National Environment Program
USGCRP	United States Global Change Research Program
UV	Ultraviolet
WCRP	World Climate Research Program
WHO	World Health Organization
WMO	World Meteorological Organization
kWh	Kilowatt hour

INTRODUCTION

A New Partnership in Stewardship

Robert E. Sievers*
General Chairperson
CHEMRAWN VII

The subject of this conference, global change, means many things to different people. Most of us tend to think first of the alterations in the chemical composition of the atmosphere. Measurements have been made of carbon dioxide, chlorofluorocarbons, methane, and other species, with sufficient precision and accuracy to leave very little doubt in anyone's mind that the concentrations have, in fact, increased significantly over time. Concentration changes are the most certain and best documented forms of global change. What is less clear and more uncertain are the consequences of these changes and what we can or should do to arrest or reduce the emissions.

Examination of these complex issues — at the interface between science and policy implications — is what distinguishes this conference. We brought together — with the financial assistance of many U.S. and multi-national corporations and governmental agencies — a distinguished group of participants from throughout the world, and, through the program of lectures and poster sessions, we heard from scientists and policy-makers from universities, private industry, government agencies, and environmental organizations. It is my hope that this conference will allow us to begin to forge a new partnership for the stewardship of our atmosphere.

*Professor of Chemistry and Director of the Cooperative Institute for Research in Environmental Sciences, University of Colorado, Boulder, CO, USA.

Why is this desirable? To understand why a partnership is needed, we should consider some of the other, broader, non-ecological forms of global change occurring in many parts of our world. We see a time of remarkable change — in Eastern Europe, the former Soviet Union, the Middle East, Africa, Asia and in the United States and elsewhere, as well. National governments are faced with so many serious crises simultaneously that the governed are increasingly less confident about the abilities of their governments and political leaders to deal with long term problems. In periods of cataclysmic change in the past, wise governmental leaders have forged partnerships with leaders of commerce, academe, and others, and have sought the active assistance of individuals and institutions that serve as constructive agents for change. I submit that national governments should now encourage those partnerships which are likely to foster long-term beneficial effects for our biosphere, and for the health and well-being of its inhabitants. And if national governments falter, because they are distracted by other problems such as Balkanization or economic difficulties, or are ineffective, then corporations and other institutions should step in to take up the slack.

What form might a partnership for atmospheric stewardship take? The partners should include major corporations, academe, environmental groups, and governmental agencies. Partnerships already exist, but these would certainly benefit from improved communication and from cooperative shared activities.

The Future Actions Committee has considered the questions of what scientific facts we can agree upon, and debated the thornier issues of what actions we should recommend. One specific recommendation that has been suggested is that our networks of monitoring stations be expanded to include sites based at or near private sector laboratories throughout the world. This would extend global coverage of present governmental and university-operated sites and bring the benefits of collaboration in studies of global trends and processes by private sector scientists and their companies. The details must still be worked out, but this is one illustration of a new partnership that should foster communication, cooperation and confidence among disparate groups.

Global Change is clearly an international problem, and we were pleased that several scientists from developing countries were able to participate in this conference. Altogether there were representatives from 49 different countries. While it has been the developed countries, such as my own, that have disproportionately contributed to the atmospheric contaminants that are causing global change, it is clear that all of us will share the problems that may result. Actions in the developing countries are critical to avoiding or mitigating some of the worst of the possible consequences of global change. How developing countries progress is as crucial to atmospheric quality stewardship as what courses the developed nations take. The developing countries may be at greatest risk in effectively accommodating possible consequences of the changing composition of the atmosphere. Consequently, it is critical that we involve the developing countries in this partnership. This conference and the preceding training workshop, on environmental measurements sponsored by the U.S. Environmental Protection Agency for 25 young scientists from 23 developing countries, represent one step in that process. This workshop was held at Rutgers University with the help of the United Nations Development Program, EPA, and others who contributed time and money. We are especially grateful to Nyle Brady and William Wilson for their leadership in this activity. The participants learned much that helped prepare them for participation in CHEMRAWN and that they will carry home with them. Global problems require global measurements and global solutions, so these young scientists will help in forging new partnerships.

This returns me to a theme that I have proposed for this conference — a new partnership for stewardship of the atmosphere — founded on our best understanding of the science of the Earth and its systems. If we can agree on what we know now, and identify the most important gaps and uncertainties in our knowledge, then institutions and organizations can have greater confidence that the policies or activities that are chosen will be well-founded, though perhaps controversial.

This partnership should involve corporations, governmental agencies, universities, and environmental advocacy groups. It should include scientists and policy-makers from developing and developed countries. National and international scientific organizations should serve as organizers and communication facilitators. We should not expect miracles from this partnership, and it may be difficult to start down this path, but even if we only develop confidence in each other and learn more about the atmosphere, we will have made real progress. And if our work is done well, our children may benefit greatly from what we have done.

CHEMRAWN VII is dedicated to the memory of Dr. Walter Orr Roberts, who began as the Chairperson of this conference but passed away last year. He was the founder of the National Center for Atmospheric Research and a member of faculty at the University of Colorado. Walt was an extraordinary scientist, a wonderful teacher, and a builder of organizations and partnerships. He influenced greatly many of us and helped shape several of the ideas that have been set forth at this meeting.

FINDINGS AND RECOMMENDATIONS

Future Actions Committee
John W. Birks*, Chairperson

Executive Summary

No place remains on Earth where human activity has not had an impact — if not directly through deforestation, cultivation of land, damming of rivers, and urbanization, then indirectly through inadvertent modification of the chemical composition of the atmosphere. Driven by an unprecedented expansion of both the human population and its activities, these changes are now occurring on a global scale and still increasing in magnitude. Because global changes may ultimately affect the habitability of Earth, they have important implications for the future well-being of humanity. Perhaps more than any other component of the Earth system, the atmosphere is a common resource, shared by all human societies and intimately involved in many life processes.

Over Antarctica, the continent least directly affected by human activities, an "ozone hole" now develops every spring with depletions of the ozone column exceeding 50%. The scientific evidence clearly implicates industrial and consumer releases of chlorofluorocarbons (CFCs) as its cause. Ozone depletion over midlatitudes has been detected at a rate of 0.2-0.8% per year for reasons that are only partially understood, while the Arctic appears to be chemically poised for possible large ozone depletions in the future. These ozone

---

*Professor of Chemistry and Fellow of the Cooperative Institute for Research in Environmental Sciences, University of Colorado, Boulder, CO, USA.

depletions in the stratosphere allow increased levels of biologically damaging ultraviolet radiation to reach the Earth's surface.

In the lower atmosphere, pollutants such as hydrocarbons, oxides of nitrogen, and sulfur dioxide have resulted in elevated concentrations of strong oxidants such as ozone and the deposition of strong acids such as nitric and sulfuric acids. These photochemical products, oxidants and acids, are much more damaging to crops and ecosystems than are the primary pollutants they are derived from. As fossil fuel use has increased, atmospheric oxidants and acidity have increased, not only in urban areas, but in rural areas as well.

An ever-increasing burden of greenhouse gases resulting from human activities is accumulating in the atmosphere. The buildup of carbon dioxide arises chiefly from the combustion of fossil fuels; that of other gases from various activities including agriculture. These increases are predicted, based on our current understanding of the climate system, to result in a significant global warming within several decades if emissions continue unchecked. The resultant climate change could cause losses of land to sea-level rise, serious dislocations in world food production, and widespread damage to natural ecosystems, including accelerated reductions in biodiversity and ecosystem services. The more rapid the accumulation of greenhouse gases in the atmosphere, the more disruptive their effects are likely to be, and the more difficult it will be for societies to adjust.

Given the interconnected nature of the Earth-atmosphere-ocean system and human dependence upon it, anthropogenic changes on a global scale require adequate responses from the world's scientific, technological, economic, and political communities. In the forthcoming United Nations Conference on the Environment and Development (UNCED), Heads-of-State and governments will discuss the challenges and opportunities connected with environmental crises and conceivable pathways leading to sustainable development of our civilization. In preparation for this conference, the International Council of Scientific Unions organized a special conference, ASCEND-21 (November 1991), devoted to the Agenda of Science for Environment and Development in the 21st Century (AGENDA-21).

Through the International Geosphere-Biosphere Program (IGBP) and World Climate Research Program (WCRP), organized by the International Council of Scientific Unions (ICSU) and the World Meteorological Organization (WMO) and operating within corresponding national programs, the world scientific community is already cooperating in global change research. Global aspects of atmospheric pollution and its consequences are fundamental topics for such programs as the International Global Atmospheric Chemistry Program (IGAC), which is one of the core projects of IGBP. In order to stimulate further efforts in understanding changes in the composition of the atmosphere induced by anthropogenic emissions, and to provide more accurate projections of future states of the atmosphere, the International Union of Pure and Applied Chemistry initiated the CHEMRAWN VII Conference.

Even though our understanding of atmospheric chemistry today, as presented in the scientific talks and poster sessions of the CHEMRAWN VII Conference, is incomplete, enough is known to justify serious concern and attention from the world community. In the case of stratospheric ozone depletion, human activity has already caused damage to the atmosphere; regulation of the chemical emissions responsible has been late in coming. Although projections of climate change resulting from accumulation of greenhouse gases are still uncertain, the scientific basis for concern is real and actions are called for now. Considering the large inertia and delayed response of the climate system, we cannot wait until a clear signal of climate change has been detected. Reduction in fossil fuel consumption, which may be economically advantageous, would reduce stresses of atmospheric oxidants and acid precipitation on crops and natural ecosystems as well. Lasting solutions to these global problems will require sustained multidisciplinary efforts by the community of nations.

Based on the current understanding of atmospheric chemistry, as summarized in this conference, and on the expertise of its individual members, the Future Actions Committee makes the following recommendations:

- **With Respect to All Aspects of Global Change Chemistry:**
    1. Recognize that all atmospheric problems are interrelated and connected with biospheric processes so that an integrated, multidisciplinary approach must be taken for their solution.
    2. No major experiments aimed at mitigating global change, that have regional or global consequences, should be undertaken without first securing broad international agreement.
    3. Research and develop the means for full social environmental costing of energy use.
    4. Develop ecological balance sheets (life cycle analyses) for comparison of different processes leading to similar end products.
    5. Apply incentives/disincentives to direct the huge innovative potential of public and, especially, private R&D organizations for the development of more energy-efficient industrial

processes and more productive, but sustainable, land use.

6. Provide incentives/disincentives and accelerate development of alternative energy technologies, especially solar and safe nuclear, subject to strict environmental safeguards.

7. Encourage corporations to continue the trend toward increased participation of environmental scientists in decision-making positions.

8. Forge a partnership between governments, industry and academia in establishing global change research priorities and programs and in formulating responsible policy.

- **With Respect to Education:**

9. Foster the education and professional development of atmospheric chemists worldwide, especially in developing countries.

10. Increase environmental literacy by encouraging environmental chemistry instruction as an important part of general education at all levels (elementary school through university).

11. Improve understanding of global change issues at the political level so that due account is taken of them in policy making.

12. Transfer experience and skills in atmospheric chemistry and monitoring techniques to developing countries through a continuing program of training workshops such as the one held in conjunction with the CHEMRAWN VII conference.

- **With Respect to Global Monitoring:**

13. Implement means of establishing adequate quality control in atmospheric measurements worldwide.

14. Encourage government and industry to cooperate with the atmospheric chemistry community in developing global inventories of emissions to the atmosphere.

15. Explore with industry the possibility of strengthening and expanding the existing international efforts to establish a high quality global monitoring network in developing countries.

    Such networks would enhance our understanding of atmospheric chemistry and global change by:

    a. Establishing chemical sources and deposition patterns for acid precipitation.

    b. Obtaining trends and variability of tropospheric ozone.

    c. Characterizing the global distribution of carbon monoxide and the oxides of nitrogen.

16. Monitor UV radiation and its effects on living organisms and their ecosystems, especially in the vicinity of the Antarctic.

17. Accelerate the development of both research and routine monitoring instruments.

- **With Respect to Stratospheric Ozone Depletion:**

    18. Maintain a vigorous scientific research agenda.

    19. Continue high priority attention to developing new substitutes and replacements for chlorofluorocarbons (CFCs) and encourage increased emphasis on recycling and recovery of CFCs, hydrogen-containing chlorofluorocarbons (HCFCs), hydrogen-containing fluorocarbons (HFCs) and bromine-containing compounds.

    20. Be advised that proposed new fleets of supersonic aircraft could result in large changes in stratospheric ozone concentrations and climate.

- **With Respect to Climate Change:**

    21. Promote international discussion and agreement about controlling future emissions of greenhouse gases.

    22. Obtain a detailed understanding of the global carbon cycle.

    23. Identify and quantify the sources and sinks of greenhouse gases and aerosols.

    24. Assign high priority to understanding and quantifying the many feedbacks involved in climate change.

    25. Quantify the effects of aerosols on climate, including both direct radiative effects and changes they induce in cloud albedo via their role as cloud condensation nuclei.

    26. Utilize available proxy records of climate change (e.g., tree rings, ocean and lake sediments, ice cores, pollen records) to obtain a better understanding of the causes of climate change in the past and to validate climate models.

- **With Respect to Oxidant Formation and Acid Precipitation in the Troposphere:**

    27. Establish regional networks for the early detection of "cleaner" air resulting from emissions control strategies.

    28. Elucidate how local emissions influence regional- and global-scale chemistry.

    29. Encourage research to achieve better understanding of acidification processes, including dry deposition, in natural

ecosystems, and their interactions with other human influences.

30. Strongly enhance research efforts to increase scientific knowledge of tropical atmospheric chemistry, including biotic interactions.

FINDINGS

Stratospheric Ozone

In the stratosphere, oxygen molecules are photodissociated by absorption of ultraviolet light from the sun to produce oxygen atoms, which combine with other oxygen molecules to form ozone. This formation of stratospheric ozone is not affected by human activity. Its destruction, being catalyzed by the oxides of hydrogen, nitrogen, chlorine, and bromine, is, however, enhanced by various human activities that increase the concentrations of these ozone-destroying gases above natural levels.

The concentration of ozone in the stratosphere is quite small; its mixing ratio does not exceed 10 parts-per-million by volume anywhere in the stratosphere, and, if brought to standard pressure and temperature would only be a 3-mm thick shell of gas enveloping the Earth. Aside from some light scattering by clouds and aerosol particles, ozone provides the only significant shield for the biosphere against ultraviolet radiation in the wavelength region beginning at 242 nm (the wavelength cutoff for shielding by oxygen) and extending to approximately 320 nm. Even in its natural state, the stratospheric ozone layer is an imperfect shield against wavelengths of light in the biologically damaging UV-B region, defined as 280-320 nm, with many species of plants and animals already being stressed by exposure to the amount of UV-B that leaks through; hence, the concern for any reduction in the integrity of the ozone layer.

The ozone distribution in the stratosphere is highly variable, being produced mostly near the equator and transported poleward. The ozone column is actually about two times larger at the poles, where until recently there have been few reactions to destroy it, than at the equator where it is both rapidly produced and destroyed. Large natural variabilities in ozone concentrations make trends difficult to detect. However, the recent observation of an "ozone hole" that forms every spring over Antarctica, with as much as 95% or more ozone loss at some altitudes and more than 50% loss in the ozone column, has been linked to the accumulation of CFCs and other anthropogenic compounds containing chlorine and bromine in the atmosphere.

These gases also are most likely responsible for a recently detected 0.2-0.8% per year downward trend in ozone at midlatitudes as well.

Preliminary action on the issue of ozone depletion by CFCs has already been taken at the international level via the Montreal Protocol (1987) and London Amendments (1990). Such action establishes a precedent for international cooperation to limit the release to the atmosphere of pollutants having global consequences. Unfortunately, this action came only after the formation of the "ozone hole" over Antarctica. The CFCs already released to the atmosphere are expected to continue to cause large Antarctic ozone losses well into the next century, and recent measurements indicate that stratospheric ozone in the Arctic and even at midlatitudes is now at risk.

Climate Change

Carbon dioxide, water vapor, and other trace gases absorb infrared radiation emitted from the surface of the Earth, thereby trapping some of this radiation and preventing it from escaping to space, with the result that our planet is some 33°C warmer than it would otherwise be. Without this "greenhouse effect" the Earth would be a frozen planet. Ice core data going back approximately 160,000 years tell us that the Earth has at times been warmer and at other times much colder, with a very high correlation between temperature and the atmospheric concentration of carbon dioxide. Through fossil fuel combustion and deforestation, we are currently increasing the atmospheric burden of carbon dioxide. The atmospheric concentration of carbon dioxide has increased from a pre-industrial value of 280 ppmv to more than 350 ppmv today and is increasing at a rate of 0.5%/yr.

Besides carbon dioxide, other greenhouse gases, including methane, nitrous oxide, CFCs, and tropospheric ozone, appear to be increasing at rates such that their cumulative contribution to greenhouse warming is expected to approximately match that of carbon dioxide. General circulation models predict an average increase in global temperature in the range 1.5-4.5°C for a doubling of carbon dioxide (or its equivalent), to be realized sometime in the middle of the 21st century. Such a rapid change in the Earth's climate could have far reaching effects. Sea level rise resulting from thermal expansion of water and melting of polar ice caps could result in flooding of lowlands where much of the human population is concentrated, and melting of permafrost could result in releases of methane gas, thereby amplifying the global warming effect. Changes in local temperatures and rainfall would most likely result in large shifts in agriculturally favorable climate zones, which may or may not be matched with appropriate soils. In many regions, the process of desertification might be intensified. A massive increase in loss of biodiversity and ecosystem disruption is possible as many species may be unable to adapt or migrate at a sufficient pace.

Predictions of climate change resulting from changes in the atmospheric content of greenhouse gases and aerosols rely on computer models of the climate system, usually of the type referred to as General Circulation Models (GCMs). These models attempt to quantify our understanding of the Earth's complex climate system. The models are, of course, only as good as the physics, chemistry and biology that they include. In recent years, chemistry and biology have been found to play increasingly important roles in climate processes. Clearly, GCMs must be improved if accurate predictions of climate change are to be made. As atmospheric chemists, we emphasize in our recommendations improvements in understanding of fundamental chemical processes that must be incorporated in climate models.

The greenhouse gases that have already accumulated in the atmosphere are principally due to past burning of fossil fuels by the now-industrialized nations, and these nations continue to be responsible for a vastly disproportionate fraction of greenhouse gas emissions. The U.S. and European Community, for example, account for only 4.6% and 6.4% of the Earth's population, respectively, but currently contribute 21% and 14% of the greenhouse gas warming potential. Stabilization of the atmospheric content of greenhouse gases will require the cooperation of both industrialized and developing countries.

Oxidant Formation/Acid Precipitation

In the troposphere, hydrocarbons and oxides of nitrogen, of both natural and anthropogenic origin, undergo photochemical reactions with the ultraviolet component of sunlight to form strong oxidants. Of these, ozone, hydrogen peroxide, and particularly the hydroxyl radical play important roles in determining the lifetimes of various chemical species released to the atmosphere. Such species may themselves be natural or anthropogenic and may serve as ozone-depleting agents, greenhouse gases, or precursors to acid rain. Thus, an understanding of the oxidizing properties of the atmosphere is needed for the reliable prediction of atmospheric residence times, which in turn determine the ozone-depleting potentials of some CFC substitutes, the global warming potentials of greenhouse gases, and the extent and severity of acid deposition. Surface-level oxidants such as ozone also have adverse effects on human health, crops, and forests.

Surface-level ozone has increased several-fold in most industrialized regions and is often elevated in rural areas as well. There are indications that upper-tropospheric ozone has increased several percent in the Northern Hemisphere over the past few decades, with a qualitative link to human-influenced nitrogen oxides and hydrocarbons, but with poorly defined trends and processes. Thus, a reliable, predictive capability is lacking for global tropospheric oxidizing efficiency and its impact on both regional and global budgets of trace gases.

Dry and wet deposition of sulfuric and nitric acids have resulted in damage to the ecology of many freshwater lakes and, in combination with atmospheric oxidants, may be a causative factor of forest decline in many regions of the world. Sulfuric acid is derived principally from combustion of sulfur-containing coal, while nitric acid is produced from oxides of nitrogen released to the atmosphere from a variety of combustion sources. Thus, both oxidant formation and acid deposition are directly tied to fossil fuel use. The impacts of these air pollutants on crops and natural ecosystems can be reduced through a combination of abatement technologies that limit emissions from sources such as power generation plants and automobiles and reductions in the use of fossil fuels as an energy source.

RECOMMENDATIONS
With Respect to All Aspects of Global Change Chemistry:

1. **Recognize that all atmospheric problems are interrelated and connected with biospheric processes so that an integrated, multidisciplinary approach must be taken for their solution.**

Global warming, ozone depletion, changes in the oxidative properties of the atmosphere, and acid deposition interact in complex and poorly understood ways. Global warming, for example, results in changes in atmospheric water content, dynamics of transport and deposition of atmospheric species, and rates of individual chemical reactions. Interestingly, an increased burden of greenhouse gases is expected to cause the stratosphere to cool, with important implications for stratospheric ozone depletion. Increases in the oxidative efficiency of the atmosphere result in changes in the rates of formation of strong acids and therefore in patterns of acid deposition. Oxidants also affect the lifetimes of CFC substitutes and hence their contributions to ozone depletion and climate change. Acid deposition and strong oxidants affect plants and their emissions to the atmosphere, etc. Thus, one cannot arrive at a thorough understanding of any one of these phenomena without considering the complex interplay of all of them.

2. **No major experiments aimed at mitigating global change, that have regional or global consequences, should be undertaken without first securing broad international agreement.**

Some major experiments designed to correct or mitigate climate change and ozone depletion have been suggested. The results of such experiments are unpredictable; they could upset existing balances and ecosystems. Clearly they should not be undertaken without the agreement of all who could be affected by them. In some cases that could mean the entire international community.

3. **Research and develop the means for full social environmental costing of energy use.**

The cost to consumers of all energy sources seldom includes the environmental cost of the use of that energy. The cost of nuclear energy, for example, has not included the cost of decontamination of facilities and safe disposal of radioactive byproducts. The cost of fossil fuel energy has not included any accounting of damage to human health, crops, and ecosystems by atmospheric oxidants and acids, or costs likely to be incurred in the future as a result of climate change. It has often been asserted that application of full social environmental costing would make relatively clean (and greenhouse neutral) energy sources, such as safe nuclear, solar, wind, geothermal and hydroelectric power, more competitive with fossil fuels. However, the means for full social environmental costing of energy use have not been developed. Clearly, an intense research effort is called for, with close collaboration of physical scientists, biologists, economists, social scientists, and policy makers to develop means for full social environmental costing of energy use.

4. **Develop ecological balance sheets (life cycle analyses) for comparison of different processes leading to similar end products.**

Again, the long-term costs of environmental degradation have seldom been taken into account when choosing between different chemical and manufacturing processes. Processes that are less costly in the short term may contribute more to ecological stress and actually be more costly when viewed from a long-term, global perspective. A scientific approach for the comparison of costs, risks and benefits of different industrial processes, taking full global ecological consequences into account, needs to be researched and developed. Again, such an effort requires the close collaboration of physical scientists, biologists, social scientists, economists, and policy makers.

5. **Apply incentives/disincentives to direct the huge innovative potential of public and, especially, private R&D organizations for the development of more energy-efficient industrial processes and more productive, but sustainable, land use.**

Decision-makers in industry are acutely aware of how government incentives such as tax breaks or disincentives such as new taxes or regulations can focus the R&D efforts of government laboratories and private businesses. The innovative potential of such organizations is a great asset that should be channeled especially to improve the efficiency of energy generation from fossil fuels and the efficiency of energy use in transportation and industrial processes (from raw materials through waste treatment and recycling of materials of high "primary energy content"). Greater efficiency of energy generation and use translates directly into reduced emissions of the greenhouse gas carbon dioxide and other pollutants contributing to oxidant formation and acid precipitation. Incentives/disincentives also should be used to encourage the development of alternative methods of agriculture that are both sustainable and more productive, while reducing emissions to the atmosphere of methane and nitrous oxide greenhouse gases. Other areas of R&D that should be stimulated include studies of the effects of different uses of land on climate change (albedo, relationships between forests and rainfall, etc.), research on more energy-efficient ways to desalinate seawater/brackish water, development of new materials for more energy-efficient transport systems (i.e., composite polymers, ceramics, etc.), and development of varieties of crop plants that can tolerate greater salt content of soil or increased weather variability.

6. **Provide incentives/disincentives and accelerate development of alternative energy technologies, especially solar and safe nuclear, subject to strict environmental safeguards.**

Development of alternative energy sources, such as solar (both passive and active), nuclear, hydroelectric, wind, biomass, and geothermal sources, to replace carbon-emitting fossil fuels should be actively encouraged. Solar and nuclear fission technologies are expected to have the greatest potential for implementation on a large scale. In order for new nuclear energy plants to be built, major advances in safety of operation and waste management are required. In addition to reducing emissions of carbon dioxide to the atmosphere, substitution of alternative energy sources would result in benefits from reductions in urban and regional oxidant formation and acid deposition. The development of new fuel additives, more fuel-efficient automobiles, and expansion of mass transit systems could lead to greater energy efficiency and reduced emissions of pollutants as well.

7. Encourage corporations to continue the trend toward increased participation of environmental scientists in decision-making positions.

Industrial ecology, the practice of designing and implementing manufacturing processes within the context of the external systems with which they interact, incorporates considerations of raw materials sources, recycling during manufacturing, minimizing process emissions, practicing energy efficiency in product design and in manufacturing, and designing for recycleability. Industrial ecology is thus an active way to integrate industrial activities into sustainable development approaches. These practices will be aided by the participation of knowledgeable environmental scientists at decision-making levels in corporations, and such corporations are likely to find that implementing these practices will turn out to be excellent business decisions for the 21st century.

8. Forge a partnership between governments, industry and academia in establishing global change research priorities and programs and in formulating responsible policy.

Abatement of atmospheric pollution and its consequent effects on the planet represent a challenge to governments, industry and academia. A tripartite partnership should make it possible to arrive at agreed-upon, soundly based conclusions and effective policies more quickly than if the parties act independently. Academia, industry and governments provide both fundamental and applied research. Industry has additional technological capabilities and needs help in modifying processes and products for the more environmentally benign future; companies also provide, both directly and via taxation, funds for research. Governments must legislate and regulate in order that environmentally responsible behavior is followed; governments also must allocate funds for research. Thus, creation of an effective partnership, in which all three constituencies influence and agree on research priorities, schedules of changed practices, and ongoing policies, should be mutually beneficial. All three groups should also collaborate in informing the general public of the issues, options and rationales for adopting the chosen policies. International scientific organizations provide one important means of establishing and fostering such a partnership.

With Respect to Education:

9. Foster the education and professional development of atmospheric chemists worldwide, especially in developing countries.

Atmospheric chemists in both developed and developing countries will play important roles in the future in elucidating the complex chemistry of the atmosphere and in educating policy makers in their respective governments. Unfortunately, the disciplinary basis of today's universities is inadequate for the education of tomorrow's atmospheric chemists. Being an interdisciplinary field, atmospheric chemistry requires the collaboration of faculty of many university departments (chemistry, physics, biology, engineering, etc.) for the development of the necessary graduate curricula.

10. Increase environmental literacy by encouraging environmental chemistry instruction as an important part of general education at all levels (elementary school through university).

In order to understand the monumental regional and global environmental problems faced by society, the public needs to be educated in the chemistry of the environment. Universities and both national and international scientific societies can contribute greatly to improved public understanding of environmental problems and cost-effective solutions by offering more courses for non-specialists in atmospheric and environmental chemistry than they traditionally have. Furthermore, in many countries scientific and environmental literacy could be enhanced by providing workshops for continuing education of teachers at the pre-university level. Scientific societies could contribute to this effort by developing and making available the necessary educational materials.

11. Improve understanding of global change issues at the political level so that due account is taken of them in policy making.

It is important that both the general public and political leaders be made fully aware of the cost/benefit aspects of the many choices of action/inaction that must be made with respect to global environmental change. A large part of the responsibility for this educational enterprise falls on scientists, who are best able to evaluate the efficacy of various approaches.

12. Transfer experience and skills in atmospheric chemistry and monitoring techniques to developing countries through a continuing program of training workshops such as the one held in conjunction with the CHEMRAWN VII conference.

Scientific societies, government agencies charged with environmental research and protection, and multinational corporations should assist in the design of training literature and by providing lecturers and equipment for such workshops. This form of technology transfer is essential to development of the necessary infrastructure in developing countries for environmental protection on both local and global scales.

With Respect to Global Monitoring:

13. Implement means of establishing adequate quality control in atmospheric measurements worldwide.

In order to make credible, scientifically based predictions about future trends of atmospheric chemistry, there needs to be international coordination of government agencies, universities, industry, institutes and scientific societies to ensure the quality and standardization of the essential observations and measurements. There are many constituencies with skills and experiences, including both those who practice the relevant sciences and those who design and make the necessary instruments. In order to gather the necessary data and have it in a properly interpretable form, such data must be of a standard nature and it must also be of a reliable quality. Many agencies, universities and industrial companies have capabilities in the quality assurance and control spheres, and perhaps a division of the International Union of Pure and Applied Chemistry (IUPAC) could, together with the IUPAC Environmental Program, play a role in establishing international standards in good laboratory practice and quality control. This initiative could provide a vehicle for the continuation of the excellent start made with the EPA workshop which immediately preceded this Conference. The experience and knowledge gained by the participants could be developed and extended to other scientists in their home countries through further involvement with agencies, universities and companies.

14. Encourage government and industry to cooperate with the atmospheric chemistry community in developing global inventories of emissions to the atmosphere.

Industry and government scientists have access to confidential and proprietary information necessary to establishing inventories of releases to the atmosphere of chemical species of all types. In combination with atmospheric measurements of species concentrations, emissions data allow the determination of atmospheric lifetimes. Knowledge of lifetimes of chemical species is necessary in order to model ozone depletion in the stratosphere, degree of climate change, and the spatial deposition of strong oxidants and acids.

15. **Explore with industry the possibility of strengthening and expanding the existing international efforts to establish a high quality global monitoring network in developing countries.**

Corporations are uniquely poised for making an important contribution to atmospheric chemistry via their international network of research and manufacturing facilities. Most developing countries have not yet created the infrastructure necessary to make high-quality atmospheric measurements; however, the necessary logistical support, technical expertise, and financial resources are already available at industrial facilities situated around the world. Furthermore, many developing countries are located in tropical and sub-tropical regions where reliable data are scarce. Industry should assist with coordination of all aspects of the existing efforts in these regions, including, for example, those of WMO, IGAC, START, ASEAN and various national agencies. It is especially important that an adequate system of quality assur- ance/quality control be implemented.

Such networks would enhance our understanding of atmospheric chemistry and global change by:

a. Establishing chemical sources and deposition patterns for acid precipitation.

Besides providing more information about the rate of oxidation, transport, and deposition of nitrogen and sulfur pollutants, global measurements of precipitation chemistry are necessary to obtain biogeochemical fluxes of these and other species. Current coverage is extremely sparse, and many of the data currently being obtained are unreliable. Wet and dry deposition measurements in developing countries are particularly lacking, with the result that very little is known about sulfur and nitrogen transport and deposition in the Southern Hemisphere. The participation of developing countries in such a network is both technically feasible and geographically desirable.

b. Obtaining trends and variability of tropospheric ozone.

Unlike stratospheric ozone, which is remotely sensed by satellite instruments, the global distribution, trends, and variability of tropospheric ozone are very poorly known. Tropospheric ozone is both an oxidant, exhibiting adverse effects on human health, crops and ecosystems, and a greenhouse gas. As the primary source of hydroxyl radicals in the atmosphere, it determines the lifetimes of most atmospheric pollutants, including hydrocarbons, carbon monoxide, sulfur dioxide, the oxides of nitrogen, and CFC substitutes. Research on techniques for ozone monitoring and the implementation of the initial phases of a global network are immediate needs.

c. **Characterizing the global distribution of the oxides of nitrogen and carbon monoxide.**

Oxides of nitrogen act as a "chemical switch" for ozone production and loss in the troposphere. Below a concentration of approximately 10 parts-per-trillion by volume, oxides of nitrogen react with hydrocarbons and sunlight to destroy ozone, while above this level the photochemical oxidation of hydrocarbons leads to ozone production. In a substantial fraction of the Northern Hemisphere it is thought that nitrogen oxide levels are very near the threshold for switching between ozone loss and ozone production, while in the Southern Hemisphere NO_x concentrations are much lower and ozone destruction dominates. In the Southern Hemisphere, the main source of ozone in the troposphere is transport from the stratosphere. Thus, predictions of oxidant formation resulting from anthropogenic releases of pollutants are extremely sensitive to assumptions made about the present global distribution of nitrogen oxides. Knowledge of the global distribution of carbon monoxide is important because of its effect on hydroxyl radical concentrations, which in turn determine the atmospheric lifetimes of most pollutant gases. Understanding of poorly known distributions must be refined with expanded global measurements.

16. **Monitor UV radiation and its effects on living organisms and their ecosystems, especially in the vicinity of the Antarctic.**

Although the ozone column is monitored worldwide by both ground-based and satellite instruments, there is no global network for monitoring UV radiation at ground level. It is important that such a network be established and that the measurements include the entire wavelength-dispersed spectrum. A UV monitoring network would provide early warning of the potential deleterious effects of increased levels of UV radiation resulting from stratospheric ozone depletion. Little is known about specific effects of increased UV radiation for most organisms other than human beings and some crops, although it is known to be potentially damaging to living organisms in general. Sensitivity to UV is also known to vary substantially among different life-forms. The implication of this variability for ecosystems is that changes in species compositions and increased extinctions are likely to result from increased exposures to UV, effects that would be compounded by rapid climate change and exposure to other human-caused disruptions such as air pollutants and acidification.

17. Accelerate the development of both research and routine monitoring instruments.

Our understanding of tropospheric chemistry is currently hampered by a lack of analytical instruments having the required sensitivity, selectivity, speed of analysis, and reliability for routine measurements of a number of key atmospheric species. In many cases, research instruments have only recently been developed and routine monitoring instruments are still virtually non-existent. For some chemical species, even adequate research instruments are yet to be developed.

With Respect to Stratospheric Ozone Depletion:

18. Maintain a vigorous scientific research agenda.

Although international action involving regulations of CFC emissions has already been taken, the problem of ozone depletion requires continued attention. The cause and effect relationship between CFCs released to the atmosphere and ozone depletion in the Antarctic stratosphere appears to be firmly established, but many details of the polar and global chemistry and atmospheric dynamics need to be further refined. For example, heterogeneous reactions, the importance of which was not appreciated until discovery of the ozone hole, constitute a whole new area of investigation with many unanswered questions. New catalytic cycles have only recently been identified and require further laboratory studies of the reactions involved. Additional field studies are required to validate the completeness of the chemistry included in computer models and to advance our understanding of the coupling of this chemistry with atmospheric dynamics. A particular emphasis is required on understanding the atmospheric roles of the substitutes being planned for CFCs. The observed trend of 0.2-0.8% per year ozone depletion at midlatitudes is of great concern because of possible deleterious effects on human health, UV damage to crops, and damage to many of the Earth's ecosystems. The cause of this ozone depletion remains unaccounted for in atmospheric models and probably requires the consideration of new processes. Understanding of the mechanism of this depletion is required for projections of the future integrity of the ozone shield.

19. Continue high priority attention to developing substitutes and replacements for CFCs and encourage increased emphasis on recycling and recovery of CFCs, HCFCs, HFCs and bromine-containing compounds.

Some interim CFC replacement compounds, the HCFCs, still contain chlorine, and, although less active ozone depletion agents and greenhouse gases than CFCs, their increased use could eventually have deleterious atmospheric effects. Better substitutes are called for and eventually may be necessary for replacing HCFCs completely. This will require continued priority on developing CFC and halon replacements. Closed-system recovery and recycling will continue to be necessary for existing compounds and their replacements.

20. Be advised that proposed new fleets of supersonic aircraft could result in large changes in stratospheric ozone concentrations and climate.

It is now firmly established that oxides of nitrogen derived indirectly from bacterial sources at the Earth's surface comprise the principal catalyst for ozone destruction in the natural stratosphere and that any anthropogenic inputs of oxides of nitrogen to the middle and upper stratosphere would result in a decrease in the ozone column. Recent studies have shown that proposed fleets of supersonic aircraft flying above about 20 km may lead to large ozone depletions and associated climate changes. On the other hand, expansions of subsonic aircraft flying in the lower stratosphere and troposphere may cause increases in ozone concentrations. Furthermore, these effects on ozone and the direct emissions of NO_x and water vapor by aircraft near the tropopause may have climatic consequences. More research attention should be paid to the problems of ozone depletion and climate change before the current fleets of aircraft are expanded significantly.

With Respect to Climate Change:

21. Promote international discussion and agreement about controlling future emissions of greenhouse gases.

Climate change, including global warming, could accelerate rapidly if countries now industrializing were to follow the same track as industrialized ones in relying heavily on fossil fuels for their future development. If they are to be persuaded to do otherwise, the present industrial countries, which created the problem in the past and are still adding to it, will have to take visible measures themselves to curb their greenhouse gas emissions. Any stabilization of the atmospheric content of greenhouse gases will require cooperation among all societies, taking account of their relative responsibilities for emissions in the past and potentials for future emissions, as well as their relative capabilities for curbing emissions.

22. Obtain a detailed understanding of the global carbon cycle.

Understanding the sources, sinks and chemical transformations of carbon is critical to predictions of future climate change resulting from emissions of carbon dioxide, methane and hydrocarbons to the atmosphere. Much of the difficulty associated with balancing the carbon budget is that of quantifying small, but highly significant, perturbations on large natural fluxes. Our understanding can be greatly increased by obtaining global coverage of measurements, developing methods for measuring fluxes in addition to concentrations of carbon compounds, making use of isotope ratio measurements, better quantifying sources and sinks, and interpreting the data obtained in terms of models that include interactions between the atmosphere, biosphere, lithosphere and hydrosphere.

23. Identify and quantify the sources and sinks of greenhouse gases and aerosols.

A better quantification of the magnitude of the trace-gas sources and sinks is required for an improved understanding of their roles in the atmosphere. These estimates are needed in order to rationalize the observed trends in the abundance of these compounds, and relative source strengths are an aid in identifying the dominant sources that would deserve high priority scrutiny in the consideration of any emission-reduction strategy. A better quantitative understanding of sinks is required to establish accurate lifetimes, which is a key factor in establishing the global-warming and ozone-depleting potentials of these species. In many cases, seasonal dependences of the emissions are required since many are from human-influenced natural emissions, such as the methane emissions from rice patties. For biogenic sources and sinks, mechanisms as well as quantities are poorly known. These need to be elucidated so that effective strategies to reduce net emissions can be developed.

24. Assign high priority to understanding and quantifying the many feedbacks involved in climate change.

The complex system of solid earth, ocean, atmosphere, cryosphere and biosphere contains numerous feedback loops, some negative (stabilizing) and some positive (destabilizing). In many cases where feedbacks have been identified, it is not even known whether the feedback is positive or negative. Some examples of important feedbacks include: 1) effect of temperature on the atmospheric content of water vapor (a greenhouse gas contributing to surface warming) and on cloud formation (which has a net cooling effect); 2) effect of temperature on ocean circulation and wind patterns, which in turn affect the uptake of carbon dioxide by the ocean; 3) effect of ice melting on surface albedo; 4) effect of UV-B radiation on phytoplankton productivity and thus uptake of carbon by the oceans; 5) effect of temperature on decomposition of organic matter in sea water to produce carbon dioxide; 6) effect of temperature on the solubility of carbon dioxide in seawater; 7) fertilization of plants by increased carbon dioxide concentration, resulting in increased uptake of carbon by the biosphere; 8) effect of temperature on rates of photosynthesis and respiration; 9) effects of changes in temperature and soil moisture content on carbon fixation and storage and on methane emissions; 10) effects of changes in extent and geographical distribution of vegetation on surface albedo; and 11) effects of temperature and sunlight on phytoplankton emissions of dimethyl sulfide and therefore on cloud condensation nuclei and cloud cover. A quantitative understanding of these and other feedbacks (some yet to be identified) is necessary if we are to make predictions about future states of the Earth's climate.

25. Quantify the effects of aerosols on climate, including both direct radiative effects and changes they induce in cloud albedo via their role as cloud condensation nuclei.

The net effect of aerosols on climate is believed to be one of cooling. It is possible that increases in aerosol loading in the atmosphere since the industrial revolution have masked to some extent warming that would have otherwise occurred due to increases in carbon dioxide and other greenhouse gases. Over the oceans, cloud formation is limited by the availability of cloud condensation nuclei. Sulfate and nitrate aerosols produced over land as a result of fossil fuel combustion and sulfate aerosols produced above oceans from dimethyl sulfide derived from marine phytoplankton are believed to be major sources of cloud condensation nuclei and, therefore, to influence cloud albedo and climate. It has been hypothesized that phytoplankton may be involved in a negative feedback loop in which aerosols derived from phytoplankton alter cloud albedo so as to stabilize climate. Clearly, a better understanding of the role of aerosols in the climate system is required for predictions of future climate change.

26. Utilize available proxy records of climate change (e.g., tree rings, ocean and lake sediments, ice cores, pollen records) to obtain a better understanding of the causes of climate change in the past and to validate climate models.

On geological time scales, climate has varied widely, and thus much can be learned about the Earth's climate system from proxy records. Climate models, which attempt to quantify our understanding of the climate system and predict future climate change, should be refined and tested by every means possible; one of the best validations is the requirement that models be able to reproduce the variability of climate that has occurred in the past.

With Respect to Oxidant Formation and Acid Precipitation in the Troposphere:

27. Establish regional networks for the early detection of "cleaner" air resulting from emissions control strategies.

Regulation of the emissions of atmospheric pollutants, in a number of countries throughout the world, is expected to lead to reduced levels of oxidants such as ozone and acid precipitation. Very large changes in pollutant emissions to the atmosphere could take place in Eastern Europe, for example. Also, in the U.S. a new Clean Air Act is expected to result in reductions in levels of atmospheric oxidants and acid deposition. Early detection of trends attributable to reduced emissions will allow improvement of tropospheric models and, in turn, provide a better scientific basis for future regulations. Early detection, however, will require networks of instruments of high precision, accuracy, and reliability that are frequently intercalibrated. Such a network currently does not exist, but is feasible with current technology. The problem of detecting changes in air quality is even more severe on the global scale. Over most regions of the globe, ozone measurements are not being made at all, thus severely limiting quantitative knowledge and predictive capabilities in atmospheric chemistry and pollution research. The same applies for other important atmospheric constituents such as carbon monoxide and the oxides of nitrogen.

28. Elucidate how local emissions influence regional- and global-scale chemistry.

Although a great deal of attention is now being paid to local and even continental-scale effects of emissions of species such as oxides of nitrogen, carbon monoxide, hydrocarbons, sulfur dioxide, heavy metals, and aerosols, very little consideration has been given to the effects of these emissions on a global scale. Since the discovery of acid precipitation, it has become abundantly clear that international boundaries are no barriers to the detrimental effects of these surface-level air pollutants. This is true for oxidant formation as well, and certainly the climatic effects of tropospheric ozone, sulfate particles and airborne particulates will play a role in global climate change. Regional-to-global scale processes should be better defined through field and modeling studies with the aim of improving models of coupled chemistry, transport and deposition.

29. **Encourage research to achieve a better understanding of acidification processes, including dry deposition and cloud-mediated acidification, in natural ecosystems and their interactions with other human influences.**

Deposition of strong acids and other pollutants occurs by both wet and dry processes, and these processes appear to have different biological effects. A special case of wet deposition in which plants come into contact with acidic fogs or clouds can be especially damaging. Efforts should be made to characterize worldwide spatial and temporal distribution of wet and dry deposition of strong acids and other pollutants with special emphasis on the role of cloud chemistry. It is important to be able to determine "critical loads," i.e., the amount of pollutant that can be accepted by the biosphere without producing long-term damage. These critical loads may be highly dependent on the mode of deposition.

30. **Strongly enhance research efforts to increase scientific knowledge of tropical atmospheric chemistry, including biotic interactions.**

The tropics and subtropics play an extremely important role in atmospheric chemistry on a global scale because of the large natural and rapidly growing anthropogenic emissions of gases and aerosols into the atmosphere in these regions. The tropics are already heavily polluted by biomass burning. In the future, increasing industrial pollution may cause major changes in atmospheric chemistry on all scales, leading, for instance, to enhanced oxidant formation and acid precipitation, with substantial negative impacts on ecosystems. In addition, deforestation and other land use changes, widely occurring in tropical regions, contribute emissions of several greenhouse gases and may contribute substantially to climate change.

FUTURE ACTIONS COMMITTEE

Chairperson

Prof. John Birks
Department of Chemistry and Biochemistry, and
Cooperative Institute for Research in Environmental Sciences
University of Colorado
Boulder, CO, USA 80309-0216

Members

Dr. Daniel L. Albritton
Director, Aeronomy Laboratory
325 Broadway, R/E/AL
Boulder, CO, USA 80303

Dr. Fred Bernthal
Deputy Director
National Science Foundation
Washington, D.C., USA 20550

Dr. Nyle Brady
United Nations Development Program
1889 F Street N.W.
Washington, D.C., USA 20006

Dr. Jack G. Calvert
Atmospheric Chemistry Division
National Center for Atmospheric Research
P.O. Box 3000
Boulder, CO, USA 80307-3000

Prof. Paul J. Crutzen
Max Planck Institüt für Chemie
Postfach 3060
Mainz, Germany

Dr. Anne H. Ehrlich
Department of Biological Sciences
Stanford University
Stanford, CA, USA 94305-2493

Dr. Keiichiro Fuwa
President
Society of Environmental Science
1-20-2-506 Naka Musashino
Tokyo 180, Japan

Dr. Mary L. Good
Senior Vice President
Allied Signal, Inc.
P.O. Box 1021R
Morristown, NJ, USA 07962-1021

Dr. Alan Hayes
Chairman
ICI Agrochemicals
Fernhurst, Haslemere
Surrey GU27 3JE, England

Dr. Bruce W. Karrh
Vice President
Safety, Health & Environ. Affairs
E.I. du Pont de Nemours & Co.
Wilmington, DE, USA 19898

Prof. Mohamed Kassas
Botany Department
Faculty of Science
Cairo University
Giza 12613, Egypt

Prof. V.A. Koptyug
Presidium of Academy of Sciences
Leninskii Prospekt 14
SU-117901 Moscow, Russia

Dr. Carl Heinrich Krauch
Board of Directors of the Hüls AG
Post Office Box 1320
D-4370 Marl, Germany

Dr. Gérard Mégie
Service d'Aeronomie CNRS
Université Paris 6
T15-E5-B102
4, Place Jussieu
75252 Paris Cedex 05, France

Prof. Mario J. Molina
Department of EAPS, 54-1312
Massachusetts Institute of Technology
Cambridge, MA, USA 02139

Dr. Lubos Nondek
Department of Chemistry
Water Research Institute
Podbabska 30
16000 Prague 6, Czechoslovakia

Dr. Rudolph Pariser
Science Director
E.I. du Pont de Nemours & Co. (Retired)
851 Old Public Road
Hockessin, DE, USA 19707

Prof. Cyril Ponnamperuma
Department of Chemistry
University of Maryland
College Park, MD, USA 20742

Dr. J. W. Maurits la Rivière
Secretary General of the
International Council of Scientific Unions
51 Boulevard de Montmorency
F-75016 Paris, France

Dr. Bryant W. Rossiter
Eastman Kodak Co. (Retired)
25662 Dillon Road
Laguna Hills, CA, USA 92653

Prof. Harold I. Schiff
York University
North York, Ontario
Canada M3J 1P3

Prof. Robert E. Sievers
Cooperative Institute for Research in Environmental Sciences
and Dept. of Chemistry and Biochemistry
University of Colorado
Boulder, CO, USA 80309-0216

Prof. Xiaoyan Tang
Center for Environmental Sciences
Beijing University
Beijing 100871, People's Republic of China

Sir Crispin Tickell
Warden
Green College
Woodstock Road
Oxford OX2 6HG, England

Dr. Dieter Wyrsch
Director of Research & Development
CIBA-GEIGY Limited
Dyestuffs & Chemicals Division
CH 4002, Basel, Switzerland

INTRODUCTORY REMARKS BY THE PAST PRESIDENT OF THE INTERNATIONAL UNION OF PURE AND APPLIED CHEMISTRY (IUPAC)

Valentin A. Koptyug*

We live in a time when humankind is beginning to understand the necessity of redefining goals and ways of furthering human development. Our civilization is rapidly approaching a global crisis connected with destruction of the environment and exhaustion of natural, unrenewable resources.

The forthcoming United Nations Conference on Environment and Development (UNCED) in Rio de Janeiro (June, 1992) will be devoted to discussions at the level of Heads-of-States or Governments of the challenges and opportunities connected with global crises and conceivable pathways leading away from a visible dead end and toward the possibility of sustainable development of our civilization.

Specification and realization of the sustainable development concept require much more scientific knowledge of the state and changes of the environment, ecologically cleaner technological innovations, deeper understanding and description of interrelations between environmental and economic values and reconsideration of the relative role of competition and cooperation. Thus, the role and responsibility of science is greatly increased.

*Past President, IUPAC, and Professor, Presidium of Academy of Sciences, Moscow, Russia

Taking this into account, Mr. Maurice Strong, Secretary General of UNCED, invited ICSU to act as principal scientific adviser to UNCED. In order to elaborate adequate recommendations, ICSU, under leadership of Prof. M.G.K. Menon, organized the special working conference ASCEND-21 — Agenda of Science for Environment and Development into the 21st Century. This conference was held November 25-29, 1991 in Vienna, Austria.

Outputs of the ASCEND-21 are:

- Analytical and forecasting documents relating to sixteen most important themes, in particular, reports on such topics as "Atmosphere and Climate," "Industry and Wastes," "Policies for Technology," etc.

- The Conference statement synthesizing the results of discussions of all themes as a set of recommendations, including proposals for follow-up activities by the international scientific community.

In concluding remarks to the ASCEND-21 Conference, Prof. Menon emphasized once more the great responsibility of scientific unions of the ICSU family for the success of humankind's efforts on the way to sustainability. One of the most ambitious integrated programs is now the ICSU/WMO interdisciplinary International Geosphere-Biosphere Program (IGBP) on Global Change, which includes the International Global Atmospheric Chemistry (IGAC) Program which has governmental support in many countries.

Of course IUPAC, as a member of the ICSU family, should share the above mentioned responsibility. We are trying to join efforts of IUPAC's Divisions and Commissions in the framework of the Chemistry and Environment Program launched by IUPAC two years ago and to stimulate collaboration of the world community of chemists in some active areas.

In this context, the Union pays special attention to the highly successful series of CHEMRAWN conferences. We may now consider the initiative of Mr. Bryant Rossiter and his colleagues in launching fifteen years ago this series of conferences as a remarkable foresight of future requirements of chemical science. He emphasized that the future well being of our planet and of humankind will be highly dependent upon the success and responsibility of chemists.

The CHEMRAWN VII Conference is devoted to one of the most important topics of UNCED — namely to the state and changes of the atmosphere, to probable consequences of those changes for humankind and for all life on planet Earth, and to possible counteractions.

On behalf of the Bureau of the International Union of Pure and Applied Chemistry and of Prof. A. J. Bard, the President of the Union, I wish all participants of CHEMRAWN VII an interesting, stimulating and creative conference. I hope that the Recommendations elaborated by the Conference will be reasonably global and at the same time concrete. The near future will demonstrate to what extent our civilization is able to develop, in this period of crisis, adequate recommendations and to convert words into real actions, leading to sustainable development based on integrated scientific, technological, economic and social approaches, with social and international solidarity.

Recommendations of the chemical community have to be thoroughly weighed as soon as possible. As was emphasized by the Executive Summary of the Bergen Conference on Sustainable Development, Science and Policy: "It is better to be roughly right now than to be precisely right later." But it seems to me desirable to continue this phrase in the following way: "Of course, if there is assurance that we are roughly right but not fully wrong."

In conclusion, I would like to emphasize the condition provided by the Organizing Committee for the CHEMRAWN VII Conference is that our work must be highly efficient and successful.

Introductory Remarks
by the Chairman of the
IUPAC CHEMRAWN Committee

Sir John Meurig Thomas*

There are a variety of reasons why I welcome the opportunity of saying a few words at the outset of this Conference. All thirteen of us, from eleven different countries representing all corners of the globe, who sit on the CHEMRAWN Committee of the International Union of Pure and Applied Chemistry have been engaged for nearly four years in preparing the way for the mounting of CHEMRAWN VII. It has been a long and often labyrinthine path, during the course of which we encountered financial, logistic and a variety of other problems, all of which in the fullness of time were surmounted, thanks to the efforts of many.

Quite early on in the evolution of the idea of convening CHEMRAWN VII, North American members of the main CHEMRAWN Committee took the lead in setting up the necessary structures needed for such a major venture. They were led in sterling fashion by Dr. Rudy Pariser. My colleagues on the CHEMRAWN Committee and I cannot exaggerate the role that Dr. Pariser has played in all this. He has combined clear-headedness, tact, diplomacy, financial responsibility, patience and good humor into a unique amalgam that has not only enhanced our respect and admiration of him but has secured for us all the prospect of a most timely and worthwhile conference.

---

*Chairperson, IUPAC CHEMRAWN Committee, Royal Institution of Great Britain, United Kingdom.

Professor Sievers, along with all his colleagues on the Organizing Committee of CHEMRAWN VII, as well as Dr. Calvert and his twenty-five colleagues from all over the world on the Program Committee, together with Professor John Birks and his twenty or so colleagues on the Future Actions Committee, have set in motion all the numerous events that are to be pursued in future years, for, as Professor Sievers has already emphasized, continuing action is absolutely essential if we are to succeed in our mission to respond responsibly to the problems already with us and to execute our stewardship and custodianship of the global environment.

While we all recognize that our deliberations must lead to future action — and we shall hear more from the Future Actions Committee in that regard — it is perhaps less widely appreciated that even prior to assembling for this event, an important Training Workshop has already taken place at Rutgers University, New Jersey.

This two-week workshop, based on two Environmental Protection Agency training courses, "Atmospheric Sampling and Measurement" and "Quality Assurance for Atmospheric Measurements," augmented by lectures on stratospheric ozone and global warming, and the modeling of photochemical smog, was attended by twenty-five people — young people, all in the early part of their scientific careers — from twenty-three Third World countries, from Argentina and Zambia, from Bangalore, Nepal and Uruguay and many other places. Their participation in this event has been greatly facilitated by the action of key members of the Third World Academy of Sciences.

Before I draw my remarks to a close, I would like briefly to make reference to three other things. First, to say how happy we in the CHEMRAWN community are to be associated as joint sponsors of this conference with the American Chemical Society, the US National Academy and the Third World Academy of Sciences. Second, to say how much personal pleasure it gives me to ruminate over the fact that it was one of my predecessors at the Royal Institution of Great Britain, the Irishman John Tyndall — Michael Faraday's friend and successor — who was the first to identify the greenhouse effect and the essence of global warming in his work (in the early 1860s) on the infrared absorption of gases such as carbon dioxide and methane. He carried out that work, using the equipment which we now display at the Royal Institution, in a basement laboratory that my students and I still use in the Davy-Faraday laboratory which is part of our building in the heart of London.

Last, but by no means least, I want to express, on behalf of all overseas visitors to this Meeting, our appreciation of the warmth of the welcome extended to us by our American colleagues. I refer to all the American people that visitors like me have encountered on visits such as this. It is over thirty years since I first came to the United States — to the land of J. Willard Gibbs and Abraham Lincoln, two of my heroes — but I still feel as stimulated and enthused when I come here now as I did when I first came to the USA in 1959. I could sense, especially when I spoke to those young people from the Third World who are visiting this country for the first time, that they, too, had been profoundly aroused by a combination of intellectual osmosis and spontaneous warm-heartedness, just as I continue to be whenever I come here. Thank you.

Plenary Lectures

AN APPROACH TO GLOBAL CLIMATE CHANGE: A U.S. PERSPECTIVE

D. Allan Bromley*

It is always a great pleasure for me to come to Baltimore, in part because one of the Associate Directors in my office — D.A. Henderson, former Dean of the School of Public Health here at Johns Hopkins University and the man most responsible for ridding the world of smallpox — still lives here and talks often about the city.

D.A. has agreed to join me and the other members of my office in supplying one of the most demanding and yet essential needs of the modern Presidency: the scientific and technical information that the President and other members of the White House senior staff need to address issues of national and international importance. Science may not be — and, indeed, usually is not — the only consideration in these policies: they have social, economic, and political dimensions that require just as much attention as their scientific aspects. But these policies cannot be adequately addressed without scientific or technical input.

Global change is a prime example of such a policy. Global change is certainly a physical phenomenon. It depends on the chemistry of the atmosphere and the interactions of the atmosphere with the oceans, with living creatures, and with the solid earth. Yet the issues surrounding global change extend in a substantial way into questions of

*Assistant to the President for Science and Technology, Office of Science and Technology Policy, Washington, D.C. USA

how we live today and how we will live in the future. The U.S. position toward global change must therefore be informed not only by the physical sciences but by many other considerations, and in particular by the social sciences. We must make common cause, all of us who study this issue, if we are to deal successfully with an issue of such complexity and scope.

My function, within the White House, is that of an honest broker for the scientific information on global change. I seek to provide the policy-making process with the best available scientific and technical input. I also see it as part of my job to provide some sense of how reliable that information is. In the past, there have been many cases of policy being based on very shaky science that was presented as scientific fact without any error bars. It is very important to provide a relevant degree of certainty.

New Information on the Chemistry of the Atmosphere

Certainly the past several months have kept me extremely busy in my role as an information broker. Over that period, there have been a number of major scientific developments that have influenced policy discussions at the highest levels, and many of these developments involve atmospheric chemistry.

At the end of October, it was announced that ground-based and satellite observations have shown significant ozone decreases in spring and summer in both the northern and southern hemispheres at middle and high latitudes. These decreases were larger in the 1980s than they were in the 1970s. The evidence is strong that these losses are due primarily to increases in chlorine and bromine levels, which are a result of releases of halocarbons — primarily chlorofluorocarbons and halons, respectively — into the atmosphere. We therefore believe that additional ozone losses during the 1990s will be comparable to those of the 1980s as chlorine and bromine levels continue to increase.

Until recently, the mechanism responsible for these losses was largely unknown. We now believe, however, that we understand the mechanisms responsible for the ozone hole over Antarctica. Ice crystals maintained in the lower stratosphere over the very cold circumpolar oceanic currents provide the surfaces for heterogeneous processes to significantly decrease the abundances of nitrogen oxides and increase the abundances of chlorine oxide radicals, which can then catalytically destroy ozone in the presence of sunlight. The Northern Hemisphere is distinct from the Southern in this regard. While the chemical composition of the Arctic atmosphere is highly perturbed by ice crystals during winter, there is no large ozone hole because the meteorological vortex breaks down before the onset of sunlight in springtime.

More recent work has indicated that the decrease of ozone in the northern latitudes may result from a change in the partitioning of chlorine, nitrogen, and hydrogen species due to reactions on the surface of sulfate aerosol particles. The eruption of Mount Pinatubo injected a huge pulse of sulfur aerosol precursors, mainly sulfur dioxide, into the stratosphere, setting the stage for a global experiment to establish the validity of this proposed mechanism.

The ozone losses at middle and high latitudes also have important implications for the climate. We have always thought that chlorofluorocarbons — a powerful greenhouse gas — warmed the atmosphere. But because of their destruction of ozone, the CFC's and other ozone-depleting chemicals have also had the effect of reducing radiative forcing of the troposphere in mid to high latitudes. In fact, new calculations suggest that, contrary to all prior supposition, when the direct and indirect effects are included, it may well be that the CFC's and the other ozone-depleting chemicals may have had no net effect whatever on the global greenhouse phenomena.

These have been some of the more surprising scientific developments of the past few months, but there have also been a number of others:

- On September 12, 1991, the Upper Atmosphere Research Satellite (UARS) was successfully launched aboard the space shuttle Discovery, and it is now functioning perfectly. The spacecraft has already measured the Antarctic ozone hole, which is deeper this year than ever before; now the hole has been both deep and extensive in four out of the past five years. UARS has also made the first global, space-based direct measurements of chlorine monoxide, a molecule that is now known to be the dominant catalyst in ozone destruction.

- The eruption of Mt. Pinatubo in the Philippines has injected sulfur dioxide into the upper atmosphere at abundances seen on Earth only about once every 50 to 100 years. Modeling of the effects of the volcano's cloud indicates that it is likely to cool the earth's temperature below what it would otherwise have been for several years. This eruption also provides us with a unique global experiment to study a wide range of atmospheric and global processes, such as precipitation patterns, sea surface temperatures, stratospheric temperatures, and of course a number of questions in atmospheric chemistry.

- A number of important new paleoclimatic records have become available, including tree ring data from Tasmania going back well over a millennium, new ice cores that will eventually extend 200,000 years into the past, and rock cores that will extend back 215 million years. Let me mention the tree ring data in somewhat more detail. They show that an anomalous warming is under way in Tasmania, perhaps influenced by greenhouse phenomena. This temperature increase exceeds any that have occurred over the past millennium, but it has not yet emerged clearly from the background variability of climate in this part of the Southern Hemisphere. Such data will be crucial in trying to detect the signature of true global change.

- Finally, it bears emphasis that we have effectively lost 25 percent of the anthropogenic component of the world's carbon balance. We know how much carbon dioxide is emitted into the atmosphere and how much remains in the atmosphere — about 50 percent — but we do not understand how much is taken up by the oceans or by the terrestrial biosphere. Measurements indicate that the oceans take up only about 25 percent of what is emitted into the atmosphere, implying that the land has to take up the other 25 percent, but we do not understand how this occurs.

These new findings, when taken together, have given us important new insights into the functioning of the earth's climatic system. But they have also demonstrated an extremely important aspect of this issue: our scientific understanding of climate change is far from certain. There is much that we still do not understand about our atmosphere and the behavior of its component gases, and significant surprises undoubtedly await us. There will be plenty to keep the individuals in this room busy for many years.

Indeed, we already have strong indications that our understanding of the effects of many tropospheric ozone precursors on the climate system is very uncertain. With better knowledge of the chemistry involved and with much more powerful computers to carry out the relevant chemical modeling, it is becoming possible for us to include higher-order chemical processes in our atmospheric models. Preliminary work suggests that we do not even know whether the net effect of the oxides of nitrogen is to increase or decrease the radiative forcing. In addition, second- and third-order processes involving methane as a greenhouse gas, while of substantial importance, may be less important than previously believed and comparable in magnitude to the first-order processes, thus perhaps reducing methane's importance somewhat. Such a reduction would be very important politically since methane is the gas through which the Third World currently makes a major contribution to the global greenhouse effect.

The Global Change Research Program

The scientific issues I have been discussing are all being studied through one of the more remarkable programs now being conducted by the federal government: the U.S. Global Change Research Program. This program, which has been put together through the Committee on Earth and Environmental Sciences of the Federal Coordinating Council for Science, Engineering, and Technology (FCCSET), has been a pioneering effort to mesh the activities of many different federal agencies and offices — 19 in all — into a coherent, integrated approach to the study of global change. The program has been extremely successful and has been a model — both nationally and internationally — of how many different federal entities can cooperate in specific areas of science and technology.

The Global Change Research Program has three primary goals:

- Establishing an integrated, comprehensive, long-term program of documenting the Earth system on a global scale.
- Conducting a program of focused studies to improve our understanding of the physical, geological, chemical, biological, and social processes that influence Earth system processes.
- And developing integrated conceptual and predictive Earth system models.

Atmospheric chemistry plays an integral role in each one of these goals. For example, one of the four integrated themes adopted by the program in Fiscal Year 1992 is better understanding of the global carbon cycle. Research on this subject involves quantifying sources and sinks of carbon, understanding the processes that control them, and examining how these processes might be altered by global change. We will need better information on atmospheric lifetimes, global warming potentials, chemical reactions in the atmosphere, and methods of monitoring emissions. This information will be essential in developing national and international policies for controlling emissions or altering the natural carbon cycle.

Fundamental questions of great importance remain unanswered. For example, we know that natural fluxes of carbon dioxide are approximately 20 times the anthropogenic ones, meaning that a 60 percent reduction in anthropogenic carbon dioxide emissions — which the IPCC has calculated will be needed to stabilize carbon dioxide levels in the atmosphere — is equivalent to about a 2 to 3 percent increase in the gas's natural sinks. Over the past years, several innovative proposals have been made about ways to sequester carbon dioxide and influence these natural fluxes. Of course, these proposals need much more study, but your community will be heavily involved in determining what might be viable and what side effects might be anticipated.

The Federal Response to Global Change

The federal government is now investing over $1 billion a year in the U.S. Global Change Research Program. But the Administration has always emphasized that research alone is not an adequate response to the possibility of global change. The possible consequences of climate change — in terms of agricultural productivity, changes in sea level, and different precipitation and storm patterns — are too great to be ignored. And the time lags in the earth climate system response to perturbations require that we anticipate rather than react to changes in that system.

As a result, we have already instituted a whole series of policies that will significantly reduce greenhouse gas emissions while at the same time having other important benefits. Let me list some of these actions to provide a sense of what is under way:

- The Clean Air Act Amendments of 1990 substantially reduce the emissions of a number of gases that are either greenhouse gases themselves or greenhouse gas precursors. For example, the act puts a permanent ceiling on sulfur dioxide emissions and gives utilities the flexibility to make reductions by any means. This market-based approach is a powerful incentive for energy-saving measures, which in reducing fossil fuel combustion will sharply reduce carbon dioxide emissions. It should be noted that there is no economically viable technology for capturing the carbon dioxide from fossil fuel combustion; to reduce carbon dioxide emissions, we must reduce the amount of fuel burned.

  It is important to emphasize that the sulfur aerosols from fossil fuel combustion, while they contribute in a major way to smog and acid raid, are removed from the lower atmosphere by that rain and thus never reach the lower stratosphere where they could participate in ozone destruction. The sulfur aerosols in the stratosphere come from volcanic eruptions, from ground-level emissions of carbonyl sulfide (OCS), and to a very small degree from the exhaust of high-flying aircraft.

- The National Energy Strategy includes a number of provisions that aggressively promote energy conservation. These include steps to increase efficiency in electricity generation and use — hence, decreased fuel consumption — encourage energy savings in residential and commercial buildings, boost the efficiency of industrial processes, and raise energy efficiency in transportation. The federal government is investing a significant amount in research and development for conservation and renewable energy technologies. Our Fiscal Year 1992 budget proposed that the R&D program for energy technology initiatives increase by 34 percent.

- The National Energy Strategy also promotes the development of nonfossil fuel energy sources, including nuclear, solar, and alternative fuels. The nuclear option, in my opinion, will prove to be a particularly important component of the future energy mix. Today, nuclear plants provide about 20 percent of America's electricity needs and make overall U.S. emissions of carbon dioxide 9 percent lower than they would otherwise be. Nuclear fission is the only technology currently available to us that can produce the large blocks of electrical energy now provided by fossil fuel plants with none of the associated greenhouse gas emissions. The National Energy Strategy includes a comprehensive strategy for nuclear energy development, including the development of new kinds of modular, intrinsically safe reactors, reform of the current licensing process, and the development of a permanent repository for nuclear waste.

- The United States is committed to the strengthened Montreal Protocol agreed to in London in June 1990. We also have implemented a phaseout schedule of the major CFC's and halons more rapid than is required by the Protocol, whereas few other countries have any implementation measures in place. While on the basis of the new scientific data, this is not affecting global greenhouse warming, it takes on even more importance in terms of protecting our ozone shield.

During the 1992 reassessment of the London revisions to the 1987 Montreal Protocol, the adequacy of current regulations on CFC's and halons will be examined, as will the potential need to regulate HCFC's.

All of the above actions influence *sources* of greenhouse gases. Regarding sinks for greenhouse gases, the United States is also taking active steps. For example, we are working with a number of developing countries to help assess the land available for reforestation and to help judge the feasibility of managing it to sequester carbon. Congress has also authorized the President's Tree Planting Initiative, which is designed to offset greenhouse gas emissions and achieve other important social benefits through tree planting; the program entails the planting of 1 billion trees per year on 1.5 million acres and the improvement of forest management practices.

Together, the policies I have listed are having a substantial effect on our emissions of greenhouse gases, and I believe that the Bush Administration has received far too little credit for them. At the same time, we have resisted efforts to place caps or mandate reductions specifically in our carbon dioxide emissions. We simply do not believe that such actions are justified at this time. As Roger Revelle wrote in the last article he published before his death:

> "The scientific base for a greenhouse warming is too uncertain to justify drastic action at this time. There is little risk in delaying policy responses to this century-old problem since there is every expectation that scientific understanding will be substantially improved within the next decade. Instead of premature and likely ineffective controls on fuel use, ... we may prefer to use the same resources — trillions of dollars, by some estimates — to increase our economic and technological resilience so that we can then apply specific remedies as necessary to reduce climate change or adapt to it."

This latter point, regarding adaptation to climate change, has recently been made by a subpanel of the National Research Council's Committee on Policy Implications of Greenhouse Warming. That subpanel, chaired by Paul Waggoner of the Connecticut Agricultural Experiment Station, reached the following conclusion:

> "So far as we can reason from the assumed gradual change in climate, their impacts will be no more severe, and adapting to them will be no more difficult, than for the range of climates already on earth and no more difficult than for other changes humanity faces."

These generally optimistic studies certainly have their critics. But they provide a welcome balance to the "sky is falling" rhetoric all too common elsewhere, and they are part of the data that I take into account in my role as an information broker.

Negotiations on a Framework Convention

The scientific information and policy responses on climate change form the backdrop for the ongoing international negotiations on this issue. Since last February the Intergovernmental Negotiating Committee (INC) on a Framework Convention on Climate Change has been meeting to establish the relationships, institutions, procedures, and funding mechanisms with which countries would cooperate in responding to climate change. Though the schedule will be tight, the United States remains committed to a convention that will be ready for signing during the June 1992 UN Conference on Environment and

Development (UNCED) in Rio de Janeiro. The international element of these negotiations reflects an important fact. No single country or limited group of countries can, in the longer term, have a significant effect on climate change through unilateral action. As the U.S. Office of Technology Assessment has pointed out, even if the United States were to reduce its emissions of carbon dioxide by 20 percent from current levels, it would represent a decline of just 2 percent in world-wide emissions of all greenhouse gases.

One very positive aspect of the negotiations thus far has been the progress made in the area of technology cooperation. Technology cooperation involves not just hardware but also techniques, practices, methodologies, know-how — the software needed to use technological hardware. Industrial countries and developing countries must work hand in hand to assess needs, share strengths, and find opportunities to work together to meet common goals.

I have long believed that we in the United States have a very important window of opportunity related to the provision of environmentally benign energy technology to the Third World. If we are proactive and provide this technology, know-how, and, where required, financial support, we stand to gain in three areas:

1. We will protect the global environment while fostering industrial development and economic growth in the Third World.

2. We will reap substantial positive political fall-out.

3. We will give American industry access to what will inevitably become a huge international market.

If, on the other hand, we wait until we either are forced into such action — as we will be — or are perceived to have been forced into it, then we will gain the first benefit to the global environment, as before, but we will unquestionably lose the latter two.

Conclusion

I believe that the political will and international solidarity exist to address all these issues in a meaningful and responsible way. The rapid international response to the threat of ozone depletion — a threat heightened by the development of the ozone hole over Antarctica and by the newly discovered reduction of ozone over the Northern Hemisphere — demonstrates that the nations of the world can, and must, effectively work together when a well-defined environmental threat has been identified.

In dealing with these threats to the global environment, we are recognizing a fundamental reality. The Earth is a unified system, and all of us depend on that system. In many ways, the dawning of that realization on a broad scale came a few days before Christmas in 1968, when the crew of Apollo 8 fired the engine of their Command Module and became the first human beings to ever leave the vicinity of the Earth and head out toward the Moon. When they looked back toward the Earth, for the first time humans saw the planet not as a curving blue horizon — which is the view from the Space Shuttle, for example — but as a majestic blue and white ball suspended quietly in the velvety blackness of space. Many of the Apollo astronauts have since spoken about the profound impact of that view. They were almost overwhelmed by the sheer beauty of our planet, by their feelings of loneliness as the earth receded into space, and by the seeming fragility of the planet when viewed, for the first time, as a whole.

Indeed, I am convinced that when our descendants in the distant future look back at the twentieth century, it will be remembered, perhaps more than anything else, for the photograph taken by the crew of the Apollo 8. The Earth is ours to share; it is ours to cherish; it is ours to destroy. We are truly in this together.

ATMOSPHERIC CHEMISTRY AND GLOBAL CHANGE: THE SCIENTIST'S VIEWPOINT

Daniel Albritton*

This overview examines global change from the perspective of science. The emphasis here is on climate change and greenhouse warming in particular. The rationale for this emphasis arises from the fact that humankind interfaces with global change either through its cause or its effects. Public policy decisions are focusing on both ends of this spectrum, with the most immediate being decisions associated with the human role in causing climate change, namely, human-influenced greenhouse warming.

This summary will address four points: (1) the scientific scope of global change — *what is the research arena?*, (2) a status report of its current scientific understanding — *what are the knowns and unknowns?*, (3) the status from the perspective of decisions — *what could all of this mean?* and (4) greenhouse-gas research of the 1990s and its relation to public policy — *what to look for next?*

I. The Global System: What is the Nature of the Science Arena?

It is instructive to think of the global system in terms of (i) forcings, (ii) physical processes, (iii) physical responses, (iv) biological processes, and (v) ecosystem responses. Namely, a variety of forcing agents, both natural and human-influenced, activate numerous physical processes that cause the global system to move to a new physical

*Director, Aeronomy Laboratory, National Oceanic and Atmospheric Administration, Boulder, CO, USA

state. This altered state in turn induces a variety of biological processes that cause changes in the world's ecosystems, both natural and managed.

The goal of scientific research on global change is to understand the role of humans in the forcing agents, to build a predictive understanding of how the planet will respond to these forcing, and to characterize the impacts that will ensue. Humans and their economic affairs enter into public-welfare decisions associated with the forcing agents and into similar decisions associated with impact that follow the physical and biological responses. As noted, we are at both ends of the spectrum. Specifically, there can be large costs associated with reducing our forcing of global change, and there can be large costs associated with coping with the impacts of global change. When both action and inaction can have high costs, the key to wise decisions is understanding.

Global-change science is focusing on three over-arching policy-relevant questions:

- How well do we understand our greenhouse-gas forcing?
- How well do we understand how the global system will respond?
- How well do we understand the impacts?

This summary will focus on the first two, with an emphasis on the first, because of CHEMRAWN VII's topic — atmospheric chemistry.

II. Knowns and Unknowns: How Well do We Currently Understand This System?

The complexity of the Earth system is clear, with its variety of forcing agents, diversity of "wheels", "cogs", and "linkages" that are the processes that make up the planetary machine, and the new physical and biological states that are reached. While we have learned much about this system, there is still much to learn. The status of the science varies considerably, ranging from "certainties" to "unknowns", which are arranged here in that order. There are both research and policy implications about what we do know and about what we do not know. The points given below are drawn from two recent reviews: IPCC (1990) and WMO/UNEP (1992).

(1) The Natural Greenhouse Effect (a "certainty")

In terms of basic physics: if a body is bathed in visible radiation, it warms up and radiates infrared energy (heat). In terms of our planet Earth, it works the same way, except the atmosphere introduces a "blanket" that traps part of that outbound radiation. It is not the common atmospheric gases — nitrogen and oxygen — that are the wool in the blanket. It is the minute levels of the gases like water vapor and carbon dioxide.

There are several key points regarding the greenhouse effect: (i) It is a natural part of the Earth. (ii) Water vapor and carbon dioxide have been part of the atmosphere for millions of years. (iii) Their presence has produced an average surface temperature of about 15 degrees Celsius. Without them, the average temperature would be about -15 degrees C, and our planet would be shrouded in ice.

Thus, there is no doubt that the greenhouse effect is real. We understand its basic principles. So, what is the problem and issue regarding the greenhouse effect? It is this: Just recently (geologically speaking), we have begun to alter it.

(2) Upward Trace Gas Trends ("highly confident observations")

In addition to carbon dioxide, there are several other atmospheric trace gases that cause the Earth to retain heat: methane, chlorofluorocarbons, nitrous oxide, and ozone in the lower atmosphere. Unlike carbon dioxide, these latter gases are chemically active, thus giving rise to a wider dimension of their research.

However, all do share a common property: their atmospheric abundances are increasing. Since the beginning of the industrial era, the carbon dioxide concentration in the atmosphere has increased 25%. Methane has doubled over the same period. Chlorofluorocarbons, a purely man-made molecule, did not exist in the atmosphere at the turn of the century and are increasing in abundance at a rate of several percent a year.

Thus, without a doubt, the atmospheric levels of several greenhouse gases are increasing. These fundamental observations are the pillars of the concern over human-influenced global warming.

(3) Trace Gas Forcing of the Troposphere/Surface System (a "high-trust calculation")

In the decade of the 1950s, the major gas causing radiative forcing was carbon dioxide, with the combined effect of all of the other gases amounting to only one-third as much. However, by the 1980s, not only had the total radiative forcing increased fourfold, but also carbon dioxide was then only about half of it. Thus, carbon dioxide is not the whole story and radiative forcing is increasing relatively rapidly. This demonstrates that, from a scientific perspective, both research and policy should consider all of the greenhouse gases and their relative contributions.

(4) Predictions of Future Planetary Responses (application of our "best tools")

To predict future changes as a result of current forcings requires a "working replica" of the global system. These are the global models that reside in large computers and are intended to be the best "replica models" of the complex system itself. While admittedly not perfect miniatures of the system, the best attempts have obviously been made to incorporate the known major processes. Examples of such processes and sub-systems are the following:

Cloud Feedback ("linkages"). Many of the parts are coupled cyclically. For example, a surface warming will cause more evaporation, leading to higher levels of cloudiness. Clouds can play two major roles, the two having opposite effects. First, as viewed from space, an increase in cloudiness can yield a whiter Earth, which reflects a larger fraction of the incoming solar energy back to space. This leads to a cooler surface. Second, as viewed from the surface, an increase in cloudiness yields more trapped outgoing heat radiation, which results in a warmer surface. Which dominates? When and Where? To represent these opposing processes properly is one of the many challenges to modeling the Earth system.

El Niño-Southern Oscillation (a "subassembly"). In the large expanses of the tropical Pacific, a major subsystem of the planet marches to its own drummer. The atmospheric circulation pattern (upward motion in the western Pacific, matched by downward motion in the eastern Pacific) waxes and wanes on approximately a 26-month cycle. During the high peaks and deep valleys of this variation, U.S. weather and habitation and fishing along the eastern coasts of the Pacific are impacted severely. Hence, considerable research is directed toward building a predictive capability for this major subsystem.

Oceanic Thermal Inertia (a "warm-up" period). The main delay in a warming of the Earth's surface is the long time that it takes to warm the oceans. The nature of that delay depends largely on how warm surface water is carried into the colder deep ocean and vice versa. Thus, the large-scale circulation patterns of the ocean, which are difficult to observe, are a key factor in the timing of a greenhouse warming.

Several future forcing scenarios are input to these models, e.g., (i) "business as usual" — increasing greenhouse gases and (ii) "bite the bullet" — decreasing greenhouse gases. Global models, as our best tools, then yield predictions for each of the scenarios. The results of those predictions include values for variables that are of human interest, e.g., temperature, rainfall, and sea level. Some predictions from the recent Intergovernmental Panel on Climate Change (IPCC) report are given here:

- Business-as-Usual: Temperature: up 1°C by 2025
                                     up 3°C by 2100
                      Sea Level: up 0.2 meter by 2030
                                     up 0.65 meter by 2100

- If severe cuts in greenhouse gas emissions were to be made, the above values would be reduced by factors of 2 - 5.

- Generally, such changes would not be accomplished smoothly because of superimposed natural variation.

- The continents would warm faster and more than the oceans.

It is quite important to note that it is currently believed that it is highly unlikely that no warming would occur from a "business-as-usual" greenhouse gas increase.

(5) Past Variations ("judgment calls")

The geological record contains the climate history of the planet. Proxy indicators like tree rings and fossils give estimates of past temperature variations. They show that, over the past 10,000 years since the last Ice Age, the planet's average temperature has varied 1 - 3°C, thereby providing a measure of the natural swings in the planet's surface temperature. For the past several hundred years, we have been "rebounding" from the "Little Ice Age", which was the most recent minimum in temperature.

Direct temperature measurements have been made for the past century and a half. These data show the details of the most recent end of the warming trend — an increase of about 0.45°C over that span. It has occurred largely in two upward jumps, the first in the 1920s and the second in the 1980s. Do we know why these temperature changes have occurred?

The range of these past natural variations can be compared to the above predictions. It is clear that if the predicted warmings were indeed to occur, they would happen faster and be larger than the natural changes that have occurred over the past 10,000 years.

The past variations can also be used as a baseline to aid in the search for the first signs that these predictions are actually occurring.

(6) What Cannot be Said ("unknowns")

The recent 0.45°C warming can neither confirm nor deny whether a human-influenced warming has occurred. Since the predicted greenhouse warming (about 0.7°C) is so similar in magnitude to unexplained natural variation, the signal does not stand out clearly from the noise. Thus, the "jury is still out" on whether a greenhouse warming has or has not occurred.

The current models are not accurate enough to predict regional climate changes. The greatest uncertainties arise from an imperfect understanding of: (i) sources and sinks of greenhouse gases (which influence the predicted forcing), (ii) clouds (which influence the predicted magnitude of the warming), (iii) oceans (which influence the predicted timing and patterns of the warming), and (iv) polar ice sheets (which influence the predicted sea level rises).

III. Perspective: What Does All of This Mean?

Two perspectives could be formulated, depending on one's point of view. These two extremes illuminate the issues involved.

(1) Action

- A 3°C warming would have huge impacts on society.
- Natural variation would be exceeded — both in rate and magnitude.
- Long trace gas residence times imply very slow reversibility.
- Some action sooner is easier than any action later.

(2) Inaction

- The global system is exceedingly complicated.
- Even the known feedbacks are only crudely represented in current models.
- Major discoveries/surprises are very likely.
- Greenhouse gases involve power production, transportation, agriculture, etc.

How can atmospheric chemistry research aid with the decisions of the 1990s that will have to balance these points?

IV. The Trace-Gas Research Arena of the Coming Decade: Particularly That of Most Value to Policy Makers?

Four areas of research will prove to be vital to providing a better understanding of our trace-gas forcing of the climate system and to addressing the questions that will be uppermost in decision-maker's minds in coming years:

(1) Trace Gas Sources and Sinks

The science/policy interaction is particularly critical for the greenhouse gases. The knowledge of the different sources and sinks of the variety of greenhouse gases ranges from excellent (industrial sources of chlorofluorocarbons) to abysmal (the role of iron in oceanic uptake of carbon dioxide). The studies needed to refine and quantify this knowledge are difficult, but are just the studies that will aid the policy question, "How much to cut emissions and how?"

(2) Relative Trace-Gas Contributions

The trace gases contribute differently to radiative forcing. The indices that have been used to characterize these relative contributions are Global Warming Potentials (GWPs), which depend largely on the radiative properties of a species and its atmospheric residence time. Many gases have not only the obvious direct contribution, but have several types of indirect contributions. An example of the latter is methane, which is itself a greenhouse gas and can create other greenhouse gases — water and tropospheric ozone — and influence the residence times of others. Decisions as to which gases should receive the emphasis in emission reductions will depend on their relative contributions, and scientists will be asked to refine these indices.

(3) Global Changes — Multiple Issues

A human perturbation can often cause more than one type of environmental issue. Examples are (i) the chlorofluorocarbons, which contribute to stratospheric ozone depletion, which could in turn reduce radiative forcing (the latter from reduced radiation from a cooler lower-stratosphere resulting from the ozone loss), and (ii) sulfur dioxide, which causes acid rain, but could partially offset global warming (the latter via the cooling associated with radiative scattering of the aerosols). It may be in the coming decade that science — atmospheric chemistry in particular — will be turned to by decision-makers to assist with the end of single-issue policies.

(4) Minimizing Rude Surprises

It is clear that there are areas where large uncertainties exist (e.g., regional climate change). It is also clear that there are phenomena that are, for practical purposes, irreversible. An example is the long lifetime of the chlorofluorocarbons, which imply that, even with our most stringent reductions in emissions, the Antarctic ozone "hole" is likely to be with us for a century. Consideration of action or inaction in the face of uncertainty must weigh the degree of reversibility associated with the phenomenon, i.e., whether we can indeed "quit the game later after we learn that we are being dealt losing hands." Lastly, it is also clear that discoveries and "surprises" will continue to occur. They can "cut both ways," that is, making the situation better than we thought it was and making the situation worse than we thought it was. Broad-scale observations of the global system — including both short-lived and long-lived species and made by both satellite-borne and ground-based instruments — are a key "early warning system," aiding both in discovering new characteristics of the system and teaching us that some of what we thought we knew is not true.

V. Epilogue: What Have We Learned From Past Environmental Issues?

Both science and policy have faced and struggled with a variety of environmental issues over the past two decades. Some lessons are emerging:

- *Regional scales merge into global scales.* Chlorofluorocarbons from Los Angeles contribute to the ozone hole over Antarctica. The pollution of North America and Europe may be altering the global oxidative capacity. The lesson is clear: it is one atmosphere.

- *Understanding the natural system is basic.* Chemical reactions on the surfaces of ice particles seem esoteric. But Antarctic ozone molecules have a different view. The natural system is what we inherited. It is what we are perturbing. The key to an accurate perturbation prediction is a well-understood basis set.

- *Sustained long-term research is of high value.* The past focus only on trying to solve the environmental issue at hand has been shown to be ineffective, e.g., the focus only on urban smog in the absence of (i) monitoring sources, (ii) characterizing fundamental processes, and (iii) understanding the chemistry of the surroundings. Research and crisis response are antonyms.

- *Both science and policy are iterative.* Science is not static; discoveries continue. Decisions can be made in steps, based on new knowledge. The Montreal Protocol on Substances That Deplete the Ozone Layer is an example: (i) an initial freeze on chlorofluorocarbon emissions based on the scientific understanding at the time, (ii) with subsequent polar ozone investigations providing evidence of new depletion, and (iii) a strengthening of the Protocol's provisions subsequently. Dialogue between science and policy is the key. CHEMRAWN VII is an example of such interactions.

References

IPCC, Climate Change: The IPCC Scientific Assessment, J.T. Houghton, G.J. Jenkins, and J.J. Ephraums, (eds.), Intergovernmental Panel on Climate Change, Cambridge University Press, 1990.

WMO/UNEP, Scientific Assessment of Ozone Depletion, D.L. Albritton and R.T. Watson, (Coordinators), WMO Ozone Monitoring Report No. 25, 1992.

Acknowledgment

Illuminating and enjoyable discussions with Jerry Mahlman, Bob Watson, and Susan Solomon on the above topics is gratefully acknowledged. The Atmospheric Chemistry Project of the Climate and Global Change Program of the National Oceanic and Atmospheric Administration has supported, in part, the author's description of policy-relevant research results to governmental groups and industry.

GLOBAL ENVIRONMENTAL CHANGE OVERVIEW: DEVELOPING COUNTRIES

A.P. Mitra*

INTRODUCTION

The problems, perceptions and activities (both scientific and technological) in the areas of global change have in the recent past been discussed in a number of regional and global forums. These include:

- Southern Hemisphere Perspectives of Global Change: Scientific issues, Research News and Proposed Activities, Mbabane, Swaziland, December 1988.

- IGBP Regional Meeting for South America, Brazil, March 1990.

- The Asian IGBP Workshop, New Delhi, February 1991.

- ASCA Workshop on Greenhouse Gases and Climate Change: An Asian Perspective, June 1991.

- Asian IGBP Follow-up Workshop, Singapore, December 1991.

- ICSU Conference ASCEND 21, November 1991.

The first five workshops, in particular, dealt with the problems, current efforts and the perceptions of the developing countries and a number of recommendations emerged. The common concern in all of these recommendations was: (i) lack of an adequate data base and an inadequate dissemination system, (ii) desire for cooperative programs within the regions on areas of common interest, and (iii) the need for capability building.

*Formerly Director General, Council of Scientific and Industrial Research, India

Several key areas identified for research in these recommendations were:

- Recognition of the South American continent as a unique paleoenvironmental and paleoclimatic data reservoir for the Southern Hemisphere,
- Proposal for a coordinated campaign for methane emissions from rice paddies in the Asian region,
- Need for building up a few climatic modeling centers with advanced computing and modeling capabilities: those currently being developed are in China and India (Asian Region) and Brazil (South America), and
- Increased efforts in impact studies relating to cyclone intensity and frequency and effects on agricultural crops.

A major concern expressed was the absence of national committees for IGBP in many of the developing countries. The IGBP Committees provide a nodal point and an umbrella support that is considered very valuable to developing countries. Developing countries with IGBP committees include: Bangladesh, Bolivia, Brazil, China (CAST), China (Acad. at Taipei), India, Jamaica, Kenya, Peru, Sri Lanka, Thailand, Venezuela, and Zimbabwe.

In this address, discussions will be centered on the following three major elements:

1. Parameters determining programs and options in the developing world,

2. Emergence of new efforts, and

3. New Biosphere-Atmosphere interactive measurements and modeling efforts.

PARAMETERS DETERMINING PROGRAMS AND OPTIONS IN THE DEVELOPING WORLD

Past Meteorological Data — Trend Analysis

People in the meteorological community in many of the developing countries are not convinced that there is positive evidence of warming in their own regions. Although such analyses are not always comprehensively done, the evidences are often confusing, particularly for countries located within the tropical regions. A few examples are given here.

The first example concerns the analysis of 90 years of temperature and precipitation data over the Indian subcontinent. The average temperature over the entire Indian subcontinent for the 90-year period from 1900-1989 is shown in Fig. 1 (Thapliyal and Kulshrestha,

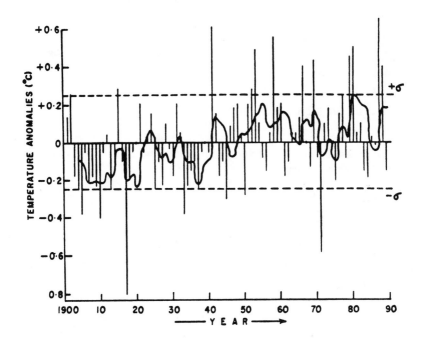

Fig. 1. Annual temperature of India. (Thick line 5-year running mean; Source: IMD)

1991). Although there is a distinct trend of an increase during this period of about 0.4°C, the fact that the changes were within the ±σ levels has cast in the minds of some meteorological scientists some doubt about the genuineness of the positive trend. The global average during this period is in the range +0.3 to +0.5°C, and the Indian results are of the same order. Analyses elsewhere give different types of results. In Vietnam, analyses of 60 years of data (1926 to 1985) of six typical meteorological stations in the two climatic regions show the following:

> North: No significant trend
> South: +0.3°C in the last two decades

In Malaysia, analyses of 11 principal meteorological stations over 100 years do show an increasing trend:

$$\overline{T}_{max} = +2.1°C$$

$$\overline{T}_{av} = +1.7°C$$

$$\overline{T}_{min} = +0.8°C$$

where $\overline{T}_{max}$ is the average daytime maximum temperature, $\overline{T}_{av}$ is the diurnal average temperature, and $\overline{T}_{min}$ is the average nighttime minimum.

There is, thus, a feeling amongst the meteorologists in the developing countries that a climate change within the tropical regions in the past century is not definitely established.

If one, however, accepts that the trend is real (as in Fig. 1 for India) then some of the results that follow are not always in consonance with results obtained elsewhere. Nevertheless, these are scientifically interesting. One such result concerns a recent analysis made over the Indian subcontinent on the trend of T_{max} and T_{min}. A remarkable result recently reported for parts of the northern hemisphere midlatitude continents is that the warming is largely characterized by increases in T_{min} rather than in T_{max}, or, in other words, the diurnal range of temperature is *decreasing* with time while the average temperature is increasing. The results over the Indian subcontinent are, however, quite different. These are shown in Fig. 2 (after Dutta, Chakravarty and Mitra, 1992). For India it is T_{max} that has largely increased while T_{min} has remained practically constant. The net result is an *increase* in the diurnal range in this region. Such regional differences are important and need further examination.

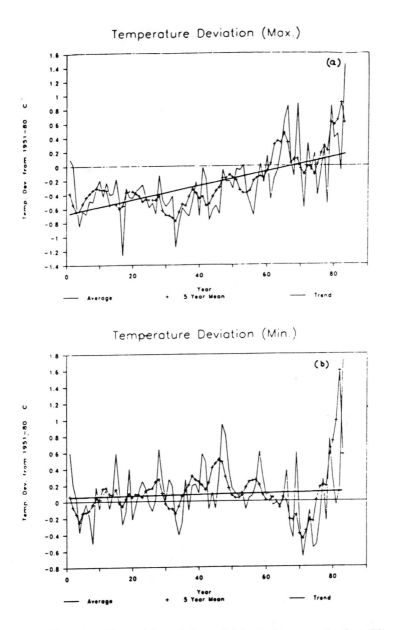

Fig. 2. Trend in T_{MAX} (a) and T_{MIN} (b) in India over the last 90 years (after Dutta, Chakravary and Mitra, 1992).

Inventory Inadequacies

Only a few of the developing countries have started making their own inventories of the greenhouse gas emissions for their countries. Most of the currently available information comes from the estimates made outside these countries on the basis primarily of statistical data available from UN organizations such as FAO and from published statistical data from these countries. In these estimates, one often resorts to extrapolation of measurements made elsewhere, assuming these are also relevant for the conditions of these countries. It must be recognized that such estimates can lead to wrong conclusions. One major example in this regard is that of methane emissions from rice paddy fields. Extrapolation of emission data from the USA and Europe to conditions in India and China had earlier given substantially larger values of methane emission than later obtained from actual measurements made in India, China and Japan.

We believe that inventories must be made by the countries themselves following international methodologies such as those of OECD (which should, however, be substantially simplified) and using emission data from our own measurements wherever available (and with intercalibration arrangements).

Only a few of the developing countries have, however, so far made serious attempts at making their own inventories. These are principally India, Brazil and China.

Part of the inadequacies arise from lack of coherence in the measurement systems (e.g., absence of calibration and standardization), inadequate networks (few and widely separated stations), and incomplete data bases. While CO_2 emissions are relatively easy to determine from existing statistical data (e.g., Oak Ridge National Laboratory CO_2 emission estimates), as also estimates of consumption and production of CFCs, estimates of CH_4 and N_2O emissions require detailed measurements of fluxes in a coordinated, intercalibrated and well standardized way. Such measurements have recently been started in India (the 1991 Paddy Field Campaign) and China. In regard to land use changes, there are sometimes discrepancies (as in the case of Brazil) between FAO estimates and those obtained from high resolution satellite data. Thus, while the satellite estimation of the average rate of deforestation in the Brazilian Amazonian Forest between 1978 and 1989 was 2.1 Mha per year, the FAO estimate for the global tropical deforestation in closed and open canopy forests for essentially the same period was 17 Mha per year.

Among the greenhouse gases, the most extensive sets of measurements in the developing countries are those of ozone. The longest series of measurements are with Dobson Spectrophotometers, but in recent years there have been additional balloonsonde measurements, as well as measurements with rockets. Dobson stations located within the latitude zone ±30° (and the dates from which measurements are available) are as follows:

NORTHERN HEMISPHERE

*1.	Quetta, Pakistan	30°11'N, 66°57'E	1957
*2.	Cairo, Egypt	30°05'N, 31°17'E	1974
*3.	New Delhi, India	28°38'N, 77°13E'	1957
4.	Naha, Japan	26°12'N, 127°41'E	1974
*5.	Varanasi, India	25°18'N, 83°01'E	1963
*6.	Kunming, China	25°01'N, 102°41'E	1980
*7.	Ahmedabad, India	23°01'N, 102°41'E	1951
	with Mt. Abu	24°36'N, 72°43'E	1969
8.	Mauna Loa, USA	19°32'N, 155°35'W	1973
9.	Mexico City, Mexico	19°20'N, 99°11'W	1973
*10.	Poona, India	18°32'N, 73°51'E	1973
*11.	Bangkok, Thailand	13°44'N, 100°34'E	1978
*12.	Kodaikanal, India	10°14'N, 77°28'E	1957
*13.	Singapore, Singapore	1°20'N, 103°53'E	1979

SOUTHERN HEMISPHERE

*14.	Huancayo, Peru	12°03'S, 75°19'W	1964
15.	Samoa, USA	14°15'S, 170°34'W	1964
16.	Cairns, Australia	16°53'S, 145°45'E	
*17.	Cachoeira Pau, Brazil	22°41'S, 45°00'W	1974
18.	Brisbane, Australia	27°25'S, 153°05'E	1957

*Stations located in developing countries.

There are thus 18 stations within ±30° latitudes and 9 within ± 20°.

For trend analysis for which a long series of data is needed, the stations that can be used are Quetta, New Delhi, Varanasi, Ahmedabad, Kodaikanal, and Huancayo.

For methane measurements, the most important network is the NOAA/CMDL cooperative flask sampling network. It should be noted that in this network there are hardly any stations from developing countries — the only exceptions are Barbados and Seychelles. There are, however, several stations in the tropical regions, and meteorological data bases are generally good and well maintained.

Special Properties of Tropical Regions

While discussing the global change activities of the developing countries, one should recognize that there are several features which are special in the tropical regions. First, the tropical regions are characterized by a relatively thin ozone column. This is seen in the average total ozone column obtained from Dobson spectrophotometers. A consequence of a thin ozone layer is a much larger dosage of UV-B in low-latitude regions. Another characteristic of the ozone column in tropical regions is that there is very little year-to-year variability. Also, the tropical atmosphere is characterized by a high tropopause.

How Developing Countries See Their Problems

To most Third World countries the concern is primarily with the *impact* aspects since a majority of these are not major contributors to the total greenhouse emissions.* The exceptions are India, Brazil and China for which the emissions are as follows (approximate values):

	India	China	Brazil
Fossil fuel (Tg CO_2–C)	130	550	53
Land use changes and biomass burning	40-80	250	400-800
Rice (Tg CH_4)	3-5	5	--
Animals (Tg CH_4)	7	--	--

The impact aspects of concern for the developing countries are:
- Human dimensions of sea-level rise,
- Changes, if any, in cyclone frequencies, flood conditions and storm surges,
- Effects on agricultural crops (especially wheat and rice), and
- UV-B effects on health (cataract problems are considered more important than skin problems).

In the ASCA meeting in Melbourne most Asian countries expressed concern about possible effects of global change on cyclone intensities and frequencies and suggested serious studies — both analysis of past observational data as well as modeling efforts — in this direction. The key element is the availability of *regional* scenarios of climate change, as opposed to global scenarios and the capability of developing countries to produce or obtain such scenarios.

Another main area of concern is the lack of an efficient communication system providing access to international data bases — a concern mentioned in all regional conferences. In this respect the Data and

Information System for the IGBP (IGBP-BIS) and the Global Climatic Observing System are of special value. The Regional Research Network program of the IGBP has major possibilities of linking activities between the developing and developed countries and also in accessing data from other centers. The program is called Global Change System for Analysis, Research and Training (START). Fourteen representative regions of START are proposed, as shown in Fig. 3.

Of highest priority, according to a meeting recently held at Bellagio in December 1990, are:

Equatorial South America:	Tropical rain forests and rapid land use changes
Northern Africa:	Extremely sensitive to climatic variation and a region of recurrent major droughts
Tropical Asian Monsoon Region:	Major global source of biogenic gases and of thermal energy to the atmosphere and a region of rapidly expanding economy

EMERGENCE OF NEW EFFORTS

In the past few years new efforts have, however, emerged in several developing countries, and in this section we outline some of these efforts. The new efforts are in the following directions:

- New measurement campaigns for greenhouse gases,
- Improvement of ozone networks and observational systems,
- Satellite remote sensing capability — IRS IA and IB satellites from India,
- ST/MST radars, and
- Inputs generated during MAP (Middle Atmosphere Program).

New Measurement Campaigns for Greenhouse Gases

A major effort has been to derive a more reliable inventory of methane emissions from paddy fields, and for this purpose large-scale measurements have been made recently in India and China and more recently in Thailand.

The Indian measurements were in the form of a campaign launched during the main rice-growing season in 1991, spanning the period August to December and involving a network of some twenty field sites distributed over the entire country and covering different soil and agricultural conditions including rainfed, waterlogged, irrigated and wetland areas. A major feature was the introduction of

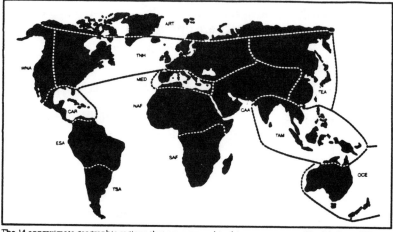

The 14 approximate geographic regions that are proposed in this report as a possible global set of RRNs. Regions and boundaries that are adopted for the global START initiative will be based on regional needs and desires, through discussions with appropriate representatives from the nations involved.

ANT	Antarctic (not shown)	OCE	Oceania
ART	Arctic	SAF	Southern and Eastern Africa
CAA	Central Arid Asia	TAM	Tropical Asian Monsoon Region
CAR	Caribbean	TEA	Temperate East Asia
ESA	Equatorial South America	TNH	Temperate Northern Hemisphere
MED	Mediterranean	TSA	Temperate South America
NAF	Northern Africa	WNA	Western North America

Fig. 3. Geographical regions proposed as a possible global set of regional research networks under IGBP.

intercalibration of all measurements through the nodal station at the National Physical Laboratory at New Delhi. Absolute calibration compatibility at the international level was established by exchanging samples with the Division of Atmospheric Physics, CSIRO, Australia and the National Institute of Agroenvironment Sciences, Japan. Scientists from Japan came to NPL, and a joint intercomparison was also made. As a result of this campaign, the total emission from India is estimated to be around 3-5 Tg CH_4 per year, less than 1/10th of the earlier estimates made elsewhere.

Another serious effort has been in estimating emissions from land use changes and biomass burning and also in providing a more accurate picture of forest areas from high resolution satellite images. Two special efforts are those relating to the Amazon Forest of Brazil and the Indian observations on the forest cover changes. For the Amazon Forest, high spectral resolution satellite data have yielded a new estimate of 2.1 Mha per year for the average rate of deforestation between 1978 and 1989 and a decrease of this value to 1.4 Mha per year in 1990 (values quite different from what one would infer from FAO figures). For India a comparison of forest cover between 1983 and 1987 shows the following:

COMPARISON OF FOREST COVER BETWEEN 1983 AND 1987
(After FRI, Dehradun)

Class	1983 position (sq kms)	1987 position (sq kms)	Net change in forest cover (sq kms)
Closed forest	361,412	378,470	+17,058 (+4.7%)
Open forest	276,583	257,409	-19,174 (-6.9%)
Mangrove forest	4,096	4,225	+179 (+4.4%)
TOTAL	642,091	640,104	-1,937 (-0.3%)

Upgrading of Ozone Network and Observational Systems

Although there has been no major increase in the number of observing stations in the developing countries, new techniques have been added in recent years. This includes the use of balloons, rockets and fairly extensive use of the satellite observations. Part of these new efforts have come from the impetus of MAP.

As an example, I will describe the new efforts in this direction made in India.

The observational techniques for India include:

- Dobson spectrophotometers at Srinagar, New Delhi, Varanasi, Mt. Abu, Pune, Kodaikanal and (for limited durations) Dakshin Gangotri (in Antarctica);

- Surface chemical ozonesondes at Pune, Trivandrum, New Delhi, Kodaikanal and Nagpur;

- Balloon measurements over Pune, Delhi, Trivandrum and Dakshin Gangotri;

- Rocket measurements at Thumba;

- Ground-based UV-B photometers at Delhi, Jodhpur, Pune and Waltair;

- A laser heterodyning system at New Delhi; and

- Microwave radiometry at 110.836 GHz (limited measurements).

The Indian Dobson network is among the densest in the tropical regions. Of 71 Dobson stations in the world, only 19 are within ±30°N,S and 9 within ±20°N,S. Of these 19, India has five.

The second major strength of the Indian ozone program is the multiplicity of techniques often used simultaneously or nearly simultaneously for profile determinations. In this context, the two rocket/balloon/ground-based intercomparison campaigns carried out at Thumba, as part of IMAP during 1983 and 1987, are of special value. A multi-technique, near-simultaneous set of profiles is given in Fig. 4.

Remote Sensing Satellites from Developing Countries

In addition to very rapid growth of remote sensing capabilities in many developing countries, a major new development has been the launching of two remote sensing satellites by India. These are the satellites IRS 1A and IRS 1B.

The IRS 1A satellite, launched on March 17, 1988, provides a major tool for Indian scientists for remote sensing of several areas of interest in global change: land use, forestry, water resources, marine resources and agricultural products. The satellite carries three cameras using charge coupled devices (CCDs) as detectors and has two types of imaging sensors — one with a spatial resolution of 72.5 meters and the other with two separate imaging sensors with spatial resolutions of 36.25 meters each.

Atmospheric Radars

A major new effort has been in the installation and operation of two ST/MST radars in Chung-li (Taiwan) and in Tirupati (India); these are

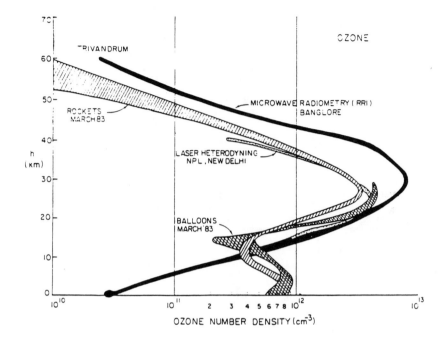

Fig. 4. Ozone profiles derived over India with different techniques: Balloonsondes, rocketsondes, laser heterodyning system, and millimeter radiometry.

in addition to the radar at Jicamarca (Peru) that has been in operation for quite some time. Some details of these three ST/MST radars are given in the following table:

GLOBAL DISTRIBUTION OF ST/MST RADARS

Facility Location	Location	Operating Frequency (MHz)	Peak Power Aperture Product Wm^2
Jicamarca, Peru	12°S, 72°W	49.9	3.2×10^{11}
Poker Flat, USA	65°N, 147°W	49.9	2.56×10^{11}
Mu, Japan	35°N, 136°E	46.5	8.33×10^9
Arecibo, Puerto Rico	19°N, 67°W	46.8	2.5×10^9
Sousy, Germany	52°N, 10°E	53.5	1.92×10^9
Tirupati, India	13°N, 79°E	53.0	3.12×10^{10}
Aberystwyth, UK	52°N, 40°W	47.0	1.25×10^9
Chung-li, Taiwan	25°N, 121°E	52.0	5×10^8
ICEAR, West Sumatra (Planned)	0°12'S, 100°E	47.0	5×10^9

The Indian radar was commissioned on ST mode on 29 October 1990 and is available to the international community of scientists. The radar provides high-resolution, 3-D contours of dynamical parameters on a real time basis. In addition, it allows a study of the troposphere characteristics with a resolution of 150 meters and consequently provides an important tool for study of stratosphere-troposphere exchange processes. The neighboring rocket range at SHAR provides opportunities for intercomparison and complementarity.

The Chung-li radar, which has been in operation for several years is a Stratospheric/Tropospheric Research Radar and Wind Profiler providing continuous radio reflectivity and three-dimensional wind profiles in the troposphere and also at the tropopause with a height resolution of 300 meters.

Another major effort concerns the planned Equatorial Radar and the International Center for Equatorial Atmospheric Research (ICEAR) to be built in west Sumatra (0°12'S, 100°19'E). It will have sensitivity of a large incoherent scatter radar and be capable of observing the entire equatorial atmosphere up to 1,000 km. It will operate on 47 MHz with an active phase array configuration similar to that of the Japanese MU radar. The planned power aperture is approximately 5×10^9 w/m^2. Thus, there is good coverage in the Asian and American sectors.

Inputs Generated During MAP

Several developing countries were major partners in the International Middle Atmosphere Program and established or upgraded a number of facilities, many of which are of direct interest to global change programs. Among the developing countries which had major programs were the People's Republic of China, Taiwan, India, Brazil, and Argentina. Rocket facilities were available both in India and Brazil. Laser radars were operational in Brazil at Sao Jose Dos Campos and in India at Trivandrum. A laser heterodyning system was established in India at New Delhi, monitoring on a real time basis profiles of ozone and water vapor. The high altitude Balloon Facility in India at Hyderabad was used extensively during this period. The current capability is fabrication and launching of balloons up to 175,000 m^3, payloads up to 100 kg, flights up to 36 km altitude, and duration of 10 to 12 hours. Joint programs with other countries were also undertaken (such as the ones between the Physical Research Laboratory in Ahmedabad and Max Plank Institute of Lindau).

A special feature of direct interest for global change was the establishment of the network of UV-B photometers to measure directly the solar UV-B radiation instead of computing it indirectly through measurements of ozone. These photometers operate at 290, 300 and 310 nm. One unit was sent to the Antarctic to measure the UV-B radiation *en route* and at the Indian Station Maitri. A contour diagram of UV-B dosages over the Indian subcontinent is shown in Fig. 5.

For the Indian program, a major target was to evolve a first order set of profiles of minor constituents, including greenhouse gases of interest (H_2O, CH_4, CFC-11, -12 and -22, and N_2O).

BIOSPHERE-ATMOSPHERE INTERACTION AND MODELING EFFORTS

Major Programs (International)

There are now several jointly arranged programs between developing and developed countries. Some of them are listed below:

Asean-Australia Project	Ocean Dynamics
UNDP-UNEP	Malaysia: Socio-economic studies Temperature and rainfall agricultural, water resources
India-Germany	Sediment Trap

Pakistan-USA   NWIO Coastal Waters 1991

Upwelling Regions   1993-1995: Research cruises by ships from
   of Arabian Sea   India, USA, Germany, the Netherlands

Indonesia-France   Ocean-atmosphere CO_2 fluxes

Greenhouse Gas Sources, Sinks and Pathways

There are now several estimates of the greenhouse gas emissions from developing countries. These estimates do not always agree, and the uncertainties are often very large.

In regard to carbon dioxide, estimates from fossil fuel inventories are available (particularly from Oak Ridge National Laboratory). Although country estimates sometimes differ from the Oak Ridge estimates (as in the case of India where country estimates are lower by about 15%) certain conclusions can be made. First, all the developing countries together contribute around 2,100 Tg per year out of the global value of 5,500, i.e. 38%. Of this amount, the main contributions come from only a few of the developing countries, as outlined below:

Asia	Latin America	Africa	Middle East
China, 554 Tg	Mexico, 74 Tg	Egypt, 20 Tg	Turkey, 34 Tg
India, 130 Tg	Brazil, 53 Tg	Algeria, 15 Tg	Iran, 31 Tg
	Argentina, 26 Tg	Nigeria, 12 Tg	Iraq, 9 Tg
	Venezuela, 26 Tg		

According to the Oak Ridge estimates, the growth rate for the developing countries is 6.4% vs. 4.4% for the developed world. However, one should note also that the per capita consumption at the present consumption and population level, is uniformly low for the developing world. While the world average is 1.2 Tg per capita, emission from India is 0.2 Tg per capita; so also are the emissions from a large number of the developing countries (Fig. 6). We have given in the inset of the diagram a scenario for policy discussions in which the increase in per capita consumption for the developing countries and the decrease from the developed countries could have a convergence value somewhere between the two, and the period needed for the convergence could be negotiated. One should note that a uniform per capita rate based on the present population is only 0.55 Tg per capita, and therefore, choice of one value for the entire world is perhaps not feasible.

One should also recognize that given the large residence times of the greenhouse gases, one is not concerned so much with emission

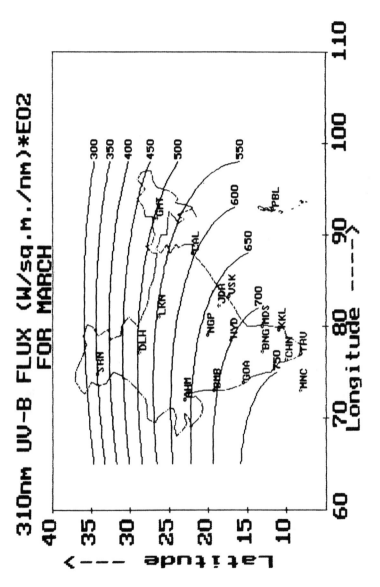

Fig. 5. Contours of 310-nm UV-B flux over the Indian subcontinent for March. Note the large changes in UV-B dosage over India (Srivastava et al., 1992).

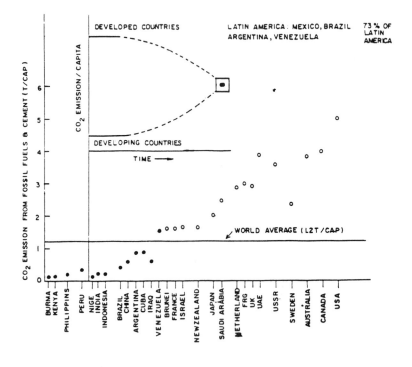

Fig. 6. CO_2 emission per capita for several developing countries vis-a-vis world average and a member of developed countries. The inset gives a possible policy scenario.

per year but more appropriately with the cumulative input, i.e. historical emissions. If one does so, then for the period 1950-1986 the cumulative emission from the USA is 27.6% (21.6% in 1986), for India 1.6% (2.6% in 1986) and for China 8% (10% in 1986).

For biomass burning, CO_2 emissions from the developing countries are somewhat large (country-wise distributions are given in Fig. 7). These are taken from Crutzen and Andreae (1990). A recent estimate in India made by the Forest Research institute gives a value around 43 Tg per year (deforestation: 0.36 Tg per year; shifting cultivation: 1.07 Tg per year; accidental fires: 27.23 Tg per year; controlled burning: 0.69 Tg per year; fire wood burning: 9.00 Tg per year). Since this is one of the major areas of concern for the developing countries, a close watch through the satellite system becomes important.

The most serious repercussions have been from the new measurement efforts of methane emission from paddy fields. Recent measurements made in India, China, Japan, Australia, and Thailand have shown emissions are significantly lower than earlier estimates and that the global value is more likely to be at the lower end of the range of 20 to 150 Tg per year mentioned in IPCC 1990 document. From the estimates made in India and information available from China, it looks very unlikely that the global estimate of methane from rice paddy fields can exceed 20-25 Tg per year. The missing sources will have to be examined.

As a case study we have examined the total greenhouse gas emissions and the consequent warming effects (using direct radiative forcing only) for India:

INDIA: CONTRIBUTION TO GLOBAL WARMING

Gas	(1)\newline Contribution to Global Warming (IPCC 90) (%)	(2)\newline Emission from India to Global (%)	(3)=(1)×(2)\newline India's Contribution to Global Warming (%)
CO_2	61	2.2 (Ours)	1.34
CH_4	15	4 (Ours)	0.6
N_2O	4	4.8 (EPA)	0.2
CFC	11	0.8	0.009
			2.23%

Role of OH in Tropical Regions

Much of the loss of methane by reaction with the hydroxyl radical (OH) in the atmosphere occurs in tropical regions, so much that it appears, based on current estimates, that the atmospheric sink *exceeds* the total sources of methane in the area defined by ±30°. This is shown in the following table as well as in Fig. 8.

TROPICAL ZONE (±30°)

Tg CH$_4$ per year

Sources		Sinks	
Rice	<80	Atmospheric Loss	300
Wetlands	30	(OH Reaction)	
Animals	40	Removal by Soil	10
Gas Venting, Leakage, and Coal Mining	20		
Land Fills	5		
Biomass Burning	30		
Termites	30		—
TOTAL	≤230	TOTAL LOSS	310

loss > emission

Analogue Approach — Use of Past Warm Periods

While there is considerable doubt about the use of past warm years as indicative of the future, they do provide high resolution temperature and precipitation scenarios and also an indication that even in regions of large scale warming, there are pockets of cooling. This type of work was made by Wigley et al. (1980) using 50 years of temperature data for the period 1925-74 and choosing 1937, 1938, 1943, 1944, and 1953 as the five warmest years and 1964, 1965, 1966, 1968 and 1972 as the five coldest years. For the Northern Hemisphere the average warming was 0.6°C, and while there was warming of varying degrees in most parts of the world, cooling was noticed over Japan, much of India and an area including and adjacent to Turkey, the Iberian Peninsula and adjacent North Africa, the Southwest coast of the USA, a region in Central Asia, and Southwestern Greenland. Also, while there was decreased precipitation over much of the USA, most of Europe, and Russia, there was an increase in precipitation over India and the Middle East. The increase in India varied from a few percent along the eastern coast (also in Bangladesh) to almost 100% in the northwest. There was, however, a decrease in Southern India and in the North around Delhi. A more recent study of the Indian subcontinent has been conducted by Dutta et al. (1992) using data of 92 stations for the period 1901 to 1983 and selecting seven colder (1903-1909) and seven warmer years (1977-1983), i.e., the first and the last decades of the total time series of the data (1901-1983).

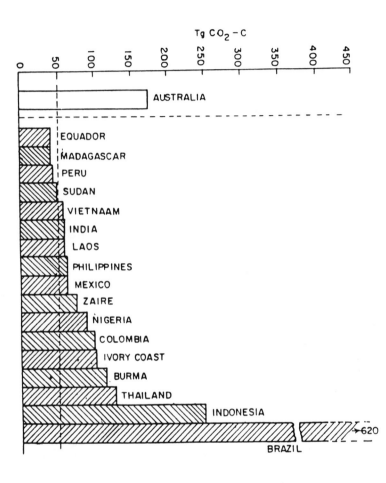

Fig. 7. CO_2 emissions from changes in forest area and associated biomass burning in tropics (after Houghton, 1991 and Houghton et al., 1987 and as tabulated in OECD 1991 — partially modified).

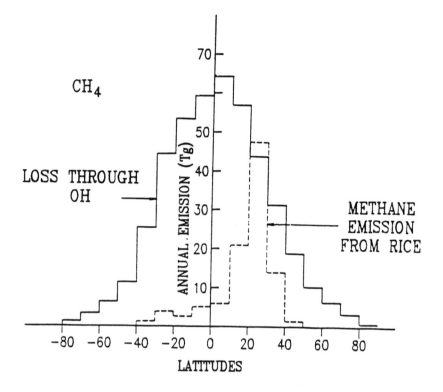

Fig. 8. Geographical distribution of methane emission from rice and atmospheric loss by reaction with OH. Note that although both are concentrated in low-latitude regions, loss to OH is considerably larger.

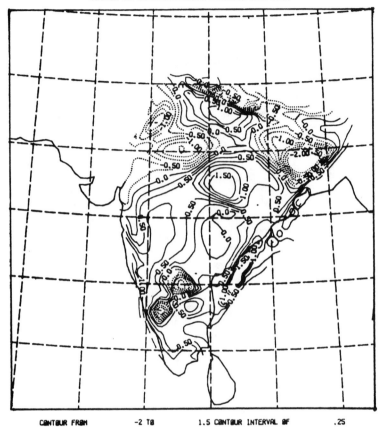

SURFACE AIR TEMPERATURE(AVR.) DEVIATION OVER INDIA

CONTOUR FROM   -2 TO   1.5 CONTOUR INTERVAL OF   .25

Fig. 9. Surface air temperature deviations over India: Warm *minus* cold years (after Dutta, Chakravary and Mitra, 1992).

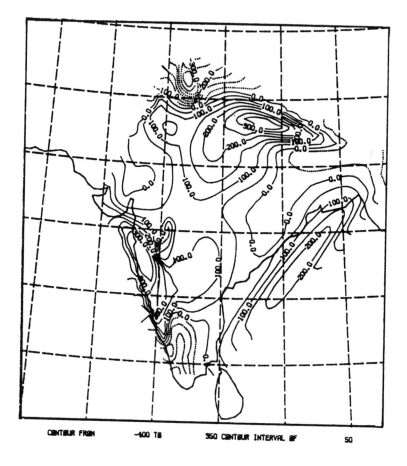

Fig. 10. Monsoon rainfall deviations over India: Warm minus cold years (after Dutta, Chakravarty and Mitra, 1992).

Differences have been obtained for the entire Indian subcontinent for T_{av}, T_{max}, and T_{min}. Except for a few localized pockets of negative deviations in T_{av}, by and large the effect is positive up to as high a value as 1.5°C (Fig. 9). Where cooling occurs, it is noted that this is mainly due to a larger decrease in T_{min} rather than in T_{max}. T_{max} changes show warming trends over almost the entire region. There is also an indication that the plains in North and Northeastern India experienced cooling while for the rest of India there is a warming, and the trend corresponds to the global rise in temperature. By and large, rainfall amount also increased with warming (Fig. 10).

Acknowledgements

I gratefully acknowledge the assistance received from Mr. R.K. Bhasin, Mrs. Sudesh Mehra and Mr. H.S. Sharma.

REFERENCES

Crutzen, P.K. and M.O. Andreas, Biomass burning in the tropics: Impact of atmospheric chemistry and biogeochemical cycles. *Science*, 250, 1669, 1990.

CSIR, Greenhouse gas emissions in India: A preliminary report, *Global Change Scientific Report No. 1.*, A.P. Mitra (ed.), CSIR, New Delhi, June 1991.

CSIR, Greenhouse gas emissions in India, *Global Change Scientific Report No. 2*, A.P. Mitra (ed.), CSIR, New Delhi, February 1992.

CSIR, CSIR Program on Global Change, New Delhi, July 1990.

DST, Geosphere Biosphere Program, Department of Science and Technology, New Delhi, February 1991.

Dutta, J., S.C. Chakravarty, and A.P. Mitra, Climate changes over the Indian subcontinent during the last 80 years, in course of publication.

Houghton, R.A., Tropical deforestation and atmospheric carbon dioxide. *Climatic Change*, in press, 1991.

Houghton, R.A., R.D. Bonne, J.R. Fruci, J.E. Hobbie, J.M. Melillo, C.A. Palm, B.J. Peterson, G.R. Shaver, and G.M. Woodwell, The flux of carbon from terrestrial ecosystems to the atmosphere in 1980 due to changes in land use: Geographic distribution of global flux. *Tellus*, 39B, 122, 1987.

IGBP, Global Change Report No. 9: Southern Hemisphere Perspectives of Global Change, B.H. Walker and R.G. Dickenson (eds.), IGBP, Stockholm.

IGBP, Global Change Report No. 12: Geosphere-Biosphere Program: A Study of Global Change, the Initial Care Projects, IGBP, Stockholm, 1990.

IGBP, Global Change Report No. 15: Global Change System for Analysis, Research and Training (START), J.A. Eddy, T.F. Malone, J.J. McCarthy, and J. Rosswall (eds.), IGBP, 1991.

IGBP, Global Change Report No. 16: Report from the IGBP Regional Meeting for South America, IGBP, Stockholm, 1991.

IGBP, Global Change Report No. 18:1, Recommendations of the Asian Workshop, New Delhi, India, 11-15 February, 1991.

IPCC, Climate Change, The IPCC Scientific Assessment, Cambridge University Press, 1990.

Khosoo, T.N., and M. Sharma (eds.), Indian Geosphere-Biosphere: Some Aspects (Proceedings of the Symposium "Indian Geosphere and Biosphere Program," held in Hyderabad, January 1990), Har-Anand Publications in association with Vikas Publishing House Pvt. Ltd., 1991.

Mathews, E., I. Fung and J. Lerner, Methane emission from rice cultivation: Geographic and seasonal distribution of cultivated areas and emissions. *Global Biogeoch. Cycles, 5*(1), 3, 1991.

Mitra, A.P., Indian Programme on Middle Atmosphere: Some results: *Adv. Space Res., 10*(10), 123, 1990.

Mitra, A.P., Overview: Global change and Indian experience, in *Impact of Global Climatic Changes on Photosynthesis and Plant Productivity,* Oxford and IBH Publishing Co., Pvt. Ltd., pp. XVII-XVIII, 1991.

N.P.L., National Workshop on ozone, B.N. Srivastava and K.S. Zalpuri (eds.), National Physical Laboratory, New Delhi, 1988.

OECD, Estimation of greenhouse gas emissions and sinks, August 1991.

Spivakovsky, C.M., S.C. Wolfsy, and M.J. Prather, A numerical method for parameterization of atmospheric chemistry: Computation of tropospheric OH. *J. Geophys. Res., 95*, D11, 18433, 1990.

Srivastava, B.N., M. Jain, M.C. Sharma, Private communication, 1992.

Stolarsky, R.S., P. Bloomfield, R. McPeter and J.R. Herman, Total ozone trends deduced from Nimbus 7 TOMS data. *Geophys. Res. Lett., 18*(6), 1015, 1991.

TERI, Proceedings of the International Conference on Global Warming and Climate Change: Perspectives from Developing Countries, S. Gupta and R.K. Pachauri (eds.), 1990.

Thapliyal, V. and Kulshrestha, S.M., Climate changes and trends over India, *Mausam*, 42(4), 333, 1991.

CONTROL OF GLOBAL CHANGE: INDUSTRIAL VIEW

E.P. Blanchard*

It is my privilege to present an industrial view of the issues surrounding atmospheric chemistry and global change. My remarks will be of a general nature, dealing with matters of policy.

Let me begin by reaffirming that industrial multinational corporations (MNCs) have the need and opportunity to address the challenges of atmospheric change. The need to act derives from a growing perception that populations and industrial activity are now at a level where they have the potential to create global climatic deviations. Two compelling forces for corporate action are the need to maintain public consent to remain in business, and the moral imperative of good stewardship.

The industrial MNCs are qualified to take a leadership role because they have many of the attributes necessary for global responses. Among these are scientific and technological resources, the means to effect technology transfer, and the opportunity to help develop and apply world standards. Perhaps most important among the characteristics of the companies is the discipline of having to combine economic, environmental and social elements in their approach.

Industry is beginning to respond in organized ways to society's demands, in cooperative efforts under the auspices of groups such as the Chemical Manufacturers Association, through the Responsible

*Vice-Chairperson, E.I. Du Pont de Nemours and Company, Wilmington, DE, USA

Care Program, and others.

Unfortunately, the demands of society are not always clear and uniform, especially when we consider global issues. For example, there are those who believe environmentalism is all about saving the planet and that in this ultimate of all crusades no trade-off is too extreme. This view ignores the reality that human beings would become extinct in a hostile environment, and the planet would go on perfectly well without us. Some other species might flourish in the new conditions. On the other hand, there are people who take a more practical view — they want to protect the environment because they understand it is essential to the quality of life as we know it.

When I was the head of the Du Pont business most involved in the chlorofluorocarbon-ozone issue, we periodically discussed the scientific data very thoroughly, as new facts became available. But I kept coming back to the question, how will all this affect future generations, including my grandchildren and their children? Inability to fully answer that question to my satisfaction weighed heavily in my decision to support CFC phaseout. Ultimately, environmentalism is a human value.

My point is, I hope we can all agree that what we are concerned about here is global change *in terms of its effects on the human condition, current and prospective.*

This is an important point because it should help determine the parameters of our efforts, and suggests broader implications than purely environmental impacts as we consider how to control global change. The question then becomes to what extent and in what circumstances is it necessary to control man's contribution to global change, consistent with maintaining or enhancing the quality of human life. Of course, we also recognize that human welfare begins at the lowest levels of the food chain, with the fate, for example, of plankton and krill.

Any discussion of the human condition must include considerations of equity. Unfortunately, in an environmental context, this inevitably creates a dichotomy. On the one hand, we must recognize that a large proportion of the people of the Earth live in extreme poverty. Their only hope for a step-change improvement in their lives is industrial development. On the other hand, such development as it was undertaken in Europe and the United States in the formative stages could not be tolerated today simply because of the numbers of people involved. Ecological systems could not survive such an onslaught in a way that would sustain the full spectrum of life support conditions as we know them. So should the affluent nations tell the people of the

Third World to suffer in silence? And would they? The answer is no, to both questions.

It is the responsibility of the global industrial community to help resolve this dilemma. The challenge is to meet the needs of future billions and to help raise up from misery those who already inhabit the impoverished regions, without bankrupting the environment. More precisely, the challenge is to generate and disseminate economic well-being without creating environmental distortions that would adversely affect the overall quality of life now and in the future.

A start has been made. Last April, in Rotterdam, 250 international companies pledged support for a set of principles embodying this concept, under the auspices of the International Chamber of Commerce and the Business Council for Sustainable Development. The group includes Du Pont, ICI, Bayer, Hoechst, BASF, Dow and other major chemical and petrochemical manufacturers.

The broader concept of human welfare is the guiding principle behind the position of Du Pont and others on global climate change. In the case of global warming, we believe there is adequate time to develop sound scientific, social, political, and economic bases for policy-making. Premature, overly restrictive actions could have disastrous effects on national economies and standards of living, as well as slowing or halting advances in technology needed for environmental progress. Therefore, public policy must be based on common understandings, developed in a global context.

Our own and other interpretations of the data indicate that some global warming is probable within the next 50 to 100 years as a result of humankind's activities. The degree of change, rate of change and regional variations still include many unanswered questions. These issues are all well-known to this audience and have been and will be explored much more rigorously during this conference. But even though the scientific jury is still out on some of the most important aspects of global warming, we know that once a climate change occurs, it may persist for decades. This is a key point in considering undesirable change — our grandchildren and great grandchildren will have to live with it.

This suggests we do not always have the luxury of waiting for conclusive evidence of deleterious effects. If we are not sure significant negative effects can be avoided, on the basis of current evidence, steps to mitigate such effects obviously must be taken as soon as is prudently possible. For example, Du Pont has committed to eliminate nitrous oxide emissions from our adipic acid plants, even before all the evidence on global warming is in, because we are aware of the long lifetime of this greenhouse gas which also apparently has ozone depleting potential.[1] We are working with other major producers to eliminate what previously had been considered an insignificant byproduct, and will meet in April to share results. The work includes field and laboratory tests, and the group is pleased by progress made to date. In addition, Du Pont is looking at a range of processes to see if there are other such byproducts that may contribute to global change.

Of course, the CFC-ozone issue brought the most important industrial response so far to environmental challenges. This issue embraces social, ethical and economic considerations of huge proportions. It involves massive investment in air-conditioning and refrigeration equipment throughout the world, as well as the capital represented in the production of CFC-based refrigerants. Compared to the investment in user equipment, the latter is relatively minor. For example, Du Pont projects spending about $1 billion to replace CFC production with alternatives. However, at a very rough estimate, we're probably talking about a quarter of a trillion dollars invested in installed equipment worldwide, for refrigeration and air conditioning. We have estimated that about $135 billion of that is in the United States.

The history of the CFC initiative is well-known, but to summarize briefly for the record, in the early 1970s some scientists theorized that CFCs might be rising into the upper atmosphere and breaking down the ozone layer. At first there was no hard evidence, but in cooperation with government and university programs Du Pont and others in industry supported further scientific research to better understand the ultimate fate of CFCs, and we concurrently began looking for alternative products. Then a significant ozone loss was detected over Antarctica. At that point, Du Pont stepped up its R&D efforts and led industry support for an international agreement — the Montreal Protocol — to limit production and use of CFCs.

In March of 1988, we saw the results of a scientifically impressive study initiated by the National Aeronautics and Space Administration, which — while still no smoking gun — led us to conclude that we must phase out these products completely. At the same time, we had responsibilities to our customers — an ethical obligation to continue to meet their needs during the transition out of CFCs. Considering that we supplied 25 percent of all CFCs used worldwide, precipitate action on our part would not have been consistent with our responsibilities to society. These include the need to maintain refrigeration for food and medical supplies, as well as for air conditioning and other uses. We decided to commit to a CFC phase-out and at the same time pull out all stops in an effort to find, develop and manufacture environmentally acceptable substitutes.

Today we have 10 plants for alternatives in operation or under construction, including two large plants — one in Ontario and one in Texas — where we have begun commercial production of a new family of alternatives. These are the Du Pont "Suva" refrigerants, announced early this year. We also offer alternatives for foam-blowing and cleaning.

Industrial users have been extremely cooperative in helping find solutions to the various challenges associated with developing workable and ozone-friendly CFC alternatives. However, there have been differences of opinion in public policy debate about which compounds are best for what purpose. This is to be expected. The transition from a few workhorse compounds to many specialized compounds will bring to life different forms of competition, the possibility of regulatory manipulation, and many opportunities for misunderstanding. There may be large investments that will fail and potentially successful ventures that will be avoided out of fear of regulatory over-control or second-guessing. But there will also be commercial successes.

The CFC initiative is certainly a watershed event in terms of international cooperation. Many valuable lessons can be learned from the protocol as we seek agreement on other aspects of climatic change. We hope the full measure of these lessons will be learned. In this I would especially include the components that led to the protocol. These are an initial scientific basis, involvement of the global community, a flexible framework treaty, fostering continued scientific input, and a role for business. I believe decisive leadership from the premier producer was essential in this case, and it also seems reasonable to suggest that only a large company with substantial scientific, technical and financial resources could have taken that leadership.

The international agreement on CFCs was unusual in its degree of success, although even this was not a complete success as several developing nations have been slow to join. One of the major obstacles to effective global responses to global problems is — to put it mildly — lack of alignment in the environmental policies of the various governments of the world. Environmental regulations and the way they are enforced vary greatly from one country or region to another. This is understandable in some cases. Obviously the priorities for a country with a shortage of food and medical supplies will be different from those in developed countries. However, the regulations also vary within the United States, where states superimpose their own rules on top of federal standards. Often environmental priorities are set by the media. This scare-of-the-month prioritizing leads to misallocation of resources and it goes without saying that it is seldom based on good science.

Ideally, there would be an objective global authority with the resources for solid scientific research, and high credibility with the public. In the United States, we have the Environmental Protection Agency, which is doing a good job but has limited resources, and of course is primarily engaged in enforcing current law. NASA, the National Science Foundation, and the National Oceanic and Atmospheric Administration are advancing our understanding of the science of climate change. Also, we have the National Laboratories, such as the one at Los Alamos, which have built up extraordinary research capability but which may have a lessening role since the end of the Cold War. The National Laboratories have the technical resources and the charter to conduct research into complex environmental issues such as global climate change, and could help strengthen the bridge between science and public policy. There will be no ideal state, but perhaps there is a vital new role here for the National Labs, and the dedicated scientists who work there.

Internationally, the best hope may well be some kind of United Nations charter that would enunciate guidelines and principles, on the premise of sustainable development. These guidelines would seek a balance between economic growth and environmental protection, along the lines of the Rotterdam Charter. We hope the Earth Charter and AGENDA-21, to emerge from next year's U.N. conference, will embody this spirit. Such guidelines could not be overly specific, of course. For example, a central planning authority could not possibly lay down specific and sensible worldwide rules regarding the desirability of incineration versus materials recycling. This would depend on location and conditions in individual cases.

The broader point here is that multinational corporations would welcome consistent international guidelines and goals, but probably would agree it would be counterproductive to include detailed rules on "how to" achieve these goals.

As S. Bruce Smart, Jr. of the World Resources Institute has pointed out, sanctions and regulations are occasionally needed, but they tend to alienate corporations and make them defensive rather than proactive.[2] He advocates that society instead should "establish the goals it wishes to achieve and translate them into a system of positive stimuli". Again we are left with the question of who should establish the goals, and create the motivation. And again we come back to the community of nations, at least partly because it already exists, in the form of the United Nations.

Most people I know in industry would be filled with apprehension at the prospect of a central bureaucracy dictating rules and regulations for the world. I share that apprehension. But at least the United Nations may be able to create some alignment among sovereign states in general principle. This would recognize the need for economic activity that increases prosperity without undermining the environment on which the economic activity depends.

The general agreement could include some ancillary commitments. For instance, no industrial plant should be built anywhere in the world without a definitive plan for waste minimization, waste treatment, and emissions reduction.

No country should subvert the principles by using them for "green protectionism." This is the practice in which a government may require very strict environmental compliance from multinationals while applying relaxed standards to its own national companies. This would in effect subsidize the country's own industry and erect a non-tariff barrier to trade. Instead, standards should be applied uniformly, and countries should seek investment by multinational companies and use it as an opportunity for the transfer of environmental technology.

In those cases where a state flagrantly disregarded the principles of environmental protection as agreed upon by the community of nations, sanctions could be employed against that state.

To return to my original premise, the role of multinational corporations in all this is multifaceted. First, we must participate in and support the development of sound science, in cooperation with other companies and institutions. There has been extraordinary cooperation in response to issues such as the effects of nitrous oxides and CFCs. But we need to be more proactive, recognizing that a profoundly deeper level of understanding is needed for predicting global change, for guiding product stewardship and for making public policy decisions. For example, Du Pont has joined the National Aeronautics and Space Administration, the National Science Foundation, Harvard, and others, in developing a high-altitude unmanned aircraft to further our knowledge of the structure of the atmosphere. A prototype of this craft, called Perseus, is currently being flight-tested over California. It may be the first of a series of increasingly sophisticated high-altitude platforms. I'm not suggesting there's a major breakthrough here — but possibly another useful tool for diagnosing atmospheric chemistry.

In summary, the global industrial enterprises are best-equipped to come up with solutions, cooperating among themselves and with institutions. A universal understanding, or philosophy, is developing around the idea of sustainable development. This is entirely consistent with the universal need to balance social, environmental, and economic considerations, and with the bottom-line discipline of the MNCs. And, finally, the MNCs can provide the mechanism for leap-frogging the technologies that lead to environmental degradation.

This would be facilitated by an international focus for policy, perhaps under the auspices of the United Nations, and an absence of governmental micro-management.

The important thing is, the industrial-scientific community must be involved in the policy-making process if for no other reason than to keep it reasonably practical. There will be no point in our complaining after the fact if the community of nations formulates policy we consider counterproductive.

REFERENCES

1. Mark H. Thiemens and William C. Trogler, *Science*, February 22, 1991.

2. NAS colloquium on industrial ecology, May, 1991.

CONTROL OF GLOBAL CLIMATE ALTERATION: POWER INDUSTRY PERSPECTIVE

George M. Hidy*

ABSTRACT

The electric utility industry is ambivalent to the issue of climate alteration. Some question the scientific interpretation of results, and others view the issue as another step towards "forced deterioration" of the U.S. energy system. Still others view actions to suppress climate alteration as a major opportunity to promote further electrification of society. Whatever the view, scientific research combined with "electro" technology development is believed to be crucial to public response. An industry R&D program has emerged that is intended to provide for an integrated assessment of the prospects for climate alteration and its environmental effects. The research complements the U.S. global climate research program. Utility interests are concentrating on environmental risk analysis opportunities by strengthening their development of options for high efficiency technologies, as well as adaptive or mitigative approaches.

I. INTRODUCTION

The purpose of this paper is to summarize the position of the electric utility industry's R&D perspective on the issue of climate alteration.

*Vice President, Electric Power Research Institute

Public awareness of the potential for anthropogenic climate alteration has increased significantly over the past decade. Man's influence on global climate is said to derive from changes in the earth's radiation balance resulting from a rising accumulation of radiatively active (greenhouse) gases (GHG) in the atmosphere. These include anthropogenic carbon dioxide (CO_2), methane (CH_4), nitrous oxide (N_2O) and chlorofluorocarbons (CFCs). The industry directly or indirectly is involved with all of these gases, through either the production or use of electricity. This industry is most often identified with CO_2 emissions, accounting for about 27% of the total U.S. emissions; the total U.S. emissions in turn are about 25% of world CO_2 emissions. As a highly regulated industry, electric utilities are an identifiable target for GHG management, even though their (U.S.) contribution to the world total atmospheric GHG burden is relatively small.

The U.S. electric utility industry is currently experiencing formidable pressures for change at both the regional and national levels. The pressures are related to increasing non-regulated competition, costs and cash flow control, fuel use and management, and multiple stresses of environmental protection. These factors, along with regional socio-political considerations, form the business environment in which utilities respond to management of environmental risks like climate alteration.

The industry response ranges from (a) sufficient skepticism about the prospects for significant climate effects to resist unwarranted changes in energy use or production to (b) a full, unqualified support for energy management minimizing GHG emissions regardless of the state of climate science. Resistance to action is partly a logical extension to utilities' responsibility for management to minimize electricity costs to the customer, while accounting for broadening requirements for all aspects of environmental protection. Alternatively, motivation for action comes from a perception that the customer (public) is aggressively supporting changes that will modernize electric services while accommodating insurance for environmental protection. Balancing these counter pressures remains the choice and judgment of each utility, given local or regional settings.

Regardless of their position, the electric utilities are very concerned about the potential for major disruption, reduction in reliability, and costs that may result if precipitous GHG emission controls were to be implemented. At the same time, the industry recognizes the potential value of "wider and wiser" use of electricity as a means to enhance world energy supply while suppressing GHG emissions. Given this setting, this paper discusses the science and mitigation issues that are reflected in the electric utility response to the risk of climate alteration. Technology alternatives for imposing efficiency on the use or generation of electricity are excluded from this paper.

II. INDUSTRY REACTIONS

A. *Questions About the Science.*

Research on climate alteration and its environmental effects continues to expand as a result of intensified interest and commitment to improved knowledge. As new information emerges, the complexities of the science have become more explicit, showing the limitations in "problem definition." While the theoretical and observational understanding of climate and its variability has improved, scientists remain at odds about the reliability of prediction of climate alteration from greenhouse gas forcing. As mathematical schemes to represent ocean interactions and cloud processes are added, for example, the apparent "sensitivity" of climate alteration estimated from current models has decreased, at least in terms of warming potential.

The potential effects of climate alteration, including precipitation patterns, occurrence and intensity of storms, disruption of ocean processes, like the current structure or the El Niño warming, remain ambiguous. Observations of surface air or sea surface temperature show well the variability in annual and interannual measures, but do not as yet confirm warming within that variability. Other measures of climate warming also yield ambiguity; for example, there is some evidence over the last century of increased occurrence of sea ice around southern Iceland, suggesting ice cap melting, while glaciers have clearly retreated in the last century. In contrast, phytoplankton activity has apparently declined slightly in the Northern Pacific (rather than increased with temperature or CO_2 absorption, as might be expected). Correlation between sunspot activity and temperature change over the last century also has resurrected the possibility of natural forcing or cycles of tens of years. These mixed results lead skeptics to question whether the theoretical predictions of climate change really represent harbingers of widespread environmental disruption.

Just as important as predicting climate in itself is the sparsity of specific information about the potential environmental effects of climate alteration. There is a list of possibilities from sea level rise or intensification of storms, to loss of water resources, agricultural productivity or natural ecosystems. For some possibilities, anecdotal information exists from historic records; for others, the effects amount to sheer speculation. Many scientists do not separate facts from opinion when speculating about environmental damage. However, the few who have actually attempted by conventional arguments to quantify damage as a function of a measure of climate alteration have found that the economic value in loss of gross world product is quite modest[1]. The valuation of possible effects using today's knowledge suggests that essentially all of the identified effects are quite manageable by technologies available today[2]. Analyses of optimization of emission management costs vs. estimated benefits of reduction of damage[3] suggest that economic damage functions would have to be highly non-linear with increasing climate alteration to stimulate political action, even into the first quarter of the next century. These considerations give focus to the strong need for greatly improved information on environmental effects so that convincing risk assessments can be done. At present, the decision makers again do not find a catastrophe in climate alteration; in one extreme, there may be no detectable effects superimposed on the "known" variability of the earth-climate response system. In this light, adaptation (as necessary) appears to be a preferable alternative to early, extensive disruption of the growing world energy supply system.

B. Vulnerability of the Energy System.

The industrial health of the United States and indeed all nations depends strongly on the reliability and robustness of the national energy supply system. The U.S. electrical supply, for example, is founded mainly on fossil fuel consumption, with about 70% using coal. Coal is also a preferred fuel for power plants for the foreseeable future, not only in the U.S. but other large countries like Russia and China. A segment of the U.S. electrical utilities fears that an early, aggressive national response shifting from a coal-based electricity production to lower CO_2 emission plants would add substantial stress to an industry already hard pressed by other issues. A search for demand-side efficiencies combined with wholesale replacement of plants, shifts to extensive use of natural gas, or addition of major emission control technology on top of sulfur and nitrogen oxide reduction requirements, could drive the industry into greatly reduced reliability in performance with a strong rise in costs. Most other nations also rely heavily on fossil fuel use, which makes the world energy system quite vulnerable to precipitous shifts in fuel resources. This vulnerability has a direct and critical importance to world economic development and public well being.

C. Opportunities for Electrification.

Perhaps the more optimistic and opportunistic segment of the industry views the issue of climate alteration as integral to a burgeoning technological change that will exploit electrification. Some have even stated that electrification worldwide represents an ultimate solution to efficient energy use, and effective energy pollution management. The applications of new end-use electro-technologies, combined with varieties of alternate generation systems and electrical transportation, are on the horizon. Realization of many environmental benefits from electrification will depend on emission controls, or alternatives for fuel use that depart from dependence on fossil fuels.

Substitution away from carbon fuels for electricity supply in the next century, of course, will be an important goal for resolving today's environmental issues. Tomorrow's issues remain to be determined, but the industry is sensitized now to search for them as an adjunct to introduction of new technologies. These are expected to include combined cycle systems with fuels cells, resurgence of nuclear power and more extensive reliance on renewables and biomass fuels, as well as solar, wind and hydro power.

III. OPPORTUNITIES FOR R&D

The electric utilities have relied heavily on four modes of research and development to maintain their technological resources. These include: (a) development programs by major equipment manufacturers and engineering firms, (b) basic and applied research through the federal government, (c) individual company or agency programs, and (d) collaborative research through institutions like the Electric Power Research Institute (EPRI). In the last decade, collaborative programs have not only become an increasingly important contributor in themselves, but also have facilitated increasingly effective cooperative programs with equipment suppliers, the government, and individual utilities.

The electric utilities have employed all of the above approaches to address the issue of global change, as a component of its broader R&D programs. Through EPRI, a major collaborative effort has been initiated that attempts to provide perspective on the range of concerns posed by global change. Programs have been established to improve our understanding of climate alteration and its environmental effects, as well as accelerate the development and demonstration of environmentally benign, efficient supply and end use technologies. Most of EPRI's program focuses on development of high efficiency and environmentally benign technologies. However, the Institute also has

initiated certain key scientific projects that are intended to comple-
ment the large, federally funded U.S. Global Change Research Program (USGCRP). EPRI's scientific work is organized to follow up the
experiences gained from earlier major, multi-disciplinary efforts like
the National Acid Precipitation Assessment Program (NAPAP).

The NAPAP exercise showed clearly that timely and concise communication of scientific results in a form useful to decision makers is
an essential format for national efforts addressing jointly environmental, energy and economic policy. Both the U.S. and international climate change research planning recognize this need, but to date have
not organized well to effectively supply (risk analysis) information to
the public (and private) sector.

A. An Integrating Framework.

Perspective on the progress in the earth sciences needs to be imbed-
ded in the advancement of understanding of the technological and
socio-economic setting that creates response to coping with global
change. An integrated assessment framework facilitates the structur-
ing of scientific results for evaluating environmental risk. Such a
framework is exemplified in Figure 1. Here the natural processes are
closely linked with feedback to the human socio-economic system to
identify conceptual important connective pathways. From such con-
ceptualizations, environmental research on global change then clearly
needs to be a balance between the earth sciences and knowledge of
human values and research. Meaningful assessment will, of course,
require such assessments on a regional scale, a scale that currently is
well beyond the capability for prediction. This balance remains to be
achieved in the USGCRP and other programs[4].

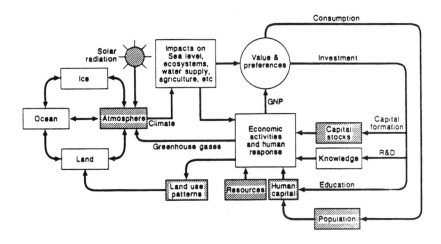

Figure 1.   Conceptual diagram illustrating the integration of Earth science research and social science research to address global change. Shading is added to topics where relatively high level of knowledge exists. (From Hidy and Peck[4])

The utility industry's research effort, as exemplified by EPRI's work, is organized to address major elements of the framework in Figure 1. This design recognizes the small utility contribution of research resources relative to the very large U.S. Global Climate Research Program.[5] Nevertheless, to date, EPRI has filled important areas of weaknesses in the federal program, including: (a) socio-economic evaluation of the impact of various measures to suppress GHG emissions, (b) the relation between costs of GHG emission reduction vs. projected environmental benefits from climate stabilization, (c) critical uncertainties in climate-general circulation models, (d) historic patterns of climate change and atmospheric chemical composition, (e) definition of potential environmental effects on utility systems, as well as natural eco-systems, and (f) development of approaches to carbon cycle modeling, incorporating human interactions with natural processes.

To strengthen the focus on the distinction between uncertainty issues that are relevant to decision making compared with emerging science, EPRI recently has joined with several agencies of the federal government to construct a conceptual framework focusing scientific studies on decision making questions. The results of this exercise should facilitate the use of an integrating framework to set research priorities for the next few years; the results from this key workshop will be available by late 1991.

B. Environmental Sciences.

Results from EPRI's studies to date in these areas have noted: (a) the potentially very high social costs of suppressing GHG emissions,[6] (b) the critical importance of improved knowledge of the significance of potential environmental effects of climate alteration for cost-benefit analysis,[3] (c) the key importance of negative feedback mechanisms (oceans and cloud systems) influencing the prediction of climate, (d) important ambiguities or inconsistencies of different climate-related records for characterizing climate change,[7] (e) the vulnerability of electricity generation to climate change,[8] (f) the specific resilience vs. vulnerability of natural ecosystems to climate change in North America,[9] and (g) critical sensitivities of the global carbon cycle to natural processes involving, for example, dissolved carbon in the oceans and soil carbon content.[10]

As scientific investigations accelerate, new results are showing the great complexity of climate forcing and feedback that underscores the high degree of uncertainty in predicting future environmental conditions. The uncertainties are compounded when investigators try to extrapolate global average forecasts to geographically specific conditions of greatest interest to decision makers. Using an integrating framework of interactive process models, risk analyses will be conducted, accounting formally for predictions in the context of cumulative uncertainties in knowledge.

Trade-offs between economic development and severe environmental stress in the regional setting will most likely motivate political actions to sacrifice the former as protection against the latter. Poorly characterized uncertainties in the knowledge leading to trade-offs will reduce confidence in the directions for societal actions. EPRI's results to date have assisted its industry sponsors and the public to appreciate better the technological and economic consequences of buying insurance to offset potential climate effects. However, EPRI's work to date, complementing the international research effort, has yielded little that would give industry leaders an ability to understand the significance of uncertainties in climate predictions to coalesce an industry response. The large uncertainties that accumulation of knowledge brings translates into results that call for great care in decision making, placing high monetary value on new information expected from the next decade of research.[3,11]

C. Mitigation Technologies.

Despite today's ambiguities in an analytical framework for deciding a course of action blending energy and environmental protection, the general direction of policy appears clear. Thus, the electric utilities need to identify and demonstrate technology options that will insure stable energy supply, with GHG emission management. In addition to developing higher efficiency generation and use technology, EPRI's program has explored opportunities for emission reductions of CO_2, or emission offsets using carbon-cycle management. In the former case, conventional carbon removal and sequestering technologies do not look economically promising for U.S. emission control. Examination of biospheric management options is currently in progress. Case studies have looked at carbon storage by forestation, by cultivation of halophyte plants, and stimulation of marine vegetation. Preliminary estimates of the upper limit capability to sequester carbon in trees annually is about 0.1-1.0 GTC (gigatons carbon) relative to current anthropogenic emissions of 5-6 GTC[12]. Widespread cultivation of halophytes in arid regions could remove roughly another 0.1-1.0 GTC annually[13]. Sequestering of carbon near shorelines by seaweed is very limited. However, stimulation of macro-algae and phytoplankton could remove larger quantities of CO_2 from the air, but would require massive management practices as yet untested at sea.

Finally, EPRI has explored the potential of aqueous, catalyzed, photoelectrochemical reactions to reduce CO_2 to hydrocarbon species.[14] At best, these reactions have similar conversion rates to photosynthesis, and do not look promising at this time for use in reducing power plant CO_2 emissions.

IV. OUTLOOK OF INTERNATIONAL UTILITIES

The electric utility industry abroad has expressed a range of views about climate alteration as in the United States. Since the industry is nationalized in many countries, its views are consistent with government policy. The international R&D agenda is similar to that of U.S. utilities. One early straw poll[15] indicated that international cooperative projects with similar objectives would have common interests among a wide range of geo-political sectors.

Studies to date have been undertaken to inform the industry about the scientific issues, while focusing on the technological opportunities for GHG management, or adaptation to climate change, should it occur.

V. SUMMARY

The electrical utility industry recognizes climate alteration as an issue for environmental protection. However, representatives of the industry are not only divided about the significance of climate change and its environmental effects, but also the extent to which energy supply strategy should account for climate alteration. Essentially all utilities, nevertheless, would subscribe to progressive efforts to improve energy use and generation efficiency. However, incorporation of GHG management in an energy strategy beyond incremental commitments is cloudy, especially when one looks at the evolution of future international growth and distribution of GHG emissions and their ability to be managed.

Electrical utilities consider that a modest research and development activity complementing the very large national programs offers an opportunity for critical insight on the issue. Added information about the relevant earth sciences, technology options, and socio-economic constraints will be essential in the short and long term for insightful decision making. In this light, utilities have invested in collaborative research aimed at supporting periodic assessments of the potential effects of climate alteration and their economic significance. At the same time, a continuing assessment of options for adaptation, mitigation and avoidance of GHG effects on climate is desirable to insure that technology is available if needed to deal with global change.

REFERENCES

1. Nordhaus, W. D., To Slow, or Not to Slow: The Economics of the Greenhouse Effect. Working Paper, Yale University, New Haven, Conn., 1990.

2. National Academy of Sciences, Policy Implications of Greenhouse Warming. National Academy Press, Washington, D.C., 1991.

3. Peck, S. C. and T. J. Teisberg, Temperature Change Related Damage Functions; A Further Analysis with CETA. Working Paper, Electric Power Research Institute, Palo Alto, CA, submitted to *Resources and Energy*, 1991.

4. Hidy, G. M. and S. Peck, Organizing for Risk Oriented Climate Alteration Research, to be published in *Air and Waste Management Assn. Journal*, December, 1991.

5. U.S. Committee on Earth and Environmental Sciences, Our Changing Planet: The FY 1992 U.S. Global Change Research Program. Office of Science and Technology Policy, Washington,

D.C., 1991.

6. Manne, A. and R. Richels, CO_2 Emission Reductions: A Regional Economic Analysis, in *Energy and Environment in the 21st Century*, MIT Press, Cambridge, 1990.

7. Falkowski, P. and C. Wilson, Phytoplankton Productivity in the North Pacific in Relation to the Absorption of Anthropogenic CO_2. Submitted to *Nature*.

8. Linder, K. P., M. J. Gibbs, and M. R. Inglis, Potential Impacts of Climate Change on Electric Utilities. Report EN-6249. Electric Power Res. Inst., Palo Alto, CA., 1989.

9. Maddox, G. D. and L. E. Morse, Large Potential Negative Effects of Global Warming on the Rare Plant Species of North America. Presented at the annual meeting of the American Society of Plant Taxonomists, August 4-8, San Antonio, TX, 1991.

10. Gherini, S., R. Hudson and R. Goldstein, A Simple, Carbon Cycle Model Linking the Atmosphere, the Hydrosphere and the Biosphere—GLOCO. In preparation, 1991.

11. Manne, A. S. and R. G. Richels, Buying Greenhouse Insurance. To be published in *Global 2100: The Economic Costs of CO_2 Emission Limits*, 1990.

12. Kulp, L., The Phytosystem as a Sink for Carbon Dioxide. Report EN-6786. Electric Power Res. Inst., Palo Alto, CA., 1990

13. Glenn, E., K. J. Kent, T. L. Thompson and R. J. Frye, Seaweeds and Halophytes to Remove Carbon from the Atmosphere. Report ER/EN-7177. Electric Power Res. Inst., Palo Alto, CA, 1991.

14. Sammells, A. F. and R. L. Cook, Direct Photoelectrochemical CO_2 Reduction to Give Useful Species. Tech. Report, Electric Power Res. Inst., Palo Alto, CA. In preparation, 1991.

15. Electric Power Research Institute, Opportunities for Collaborative Greenhouse Gas Research by the Electric Utility Industry. Report EN-7256. Electric Power Research Institute, Palo Alto, CA., 1991.

FUTURE ENERGY SOURCES AND THE ATMOSPHERIC CHALLENGES FOR RESEARCH AND DEVELOPMENT

C. H. Krauch*

What kind of energy policy will minimize global environmental damage in the next century? A challenge for research.

Summary

Reserves of fossil fuels will last a long time yet. Liquid, gaseous and solid fuels are not expected to be in short supply for at least 150 years. Liquid propellants are being produced by new methods. At a later stage the main burden of energy demand may be met by improved, safe methods of nuclear power generation. Research is still needed here.

The damage to the environment associated with the energy supply industry, and indeed with life itself, will become a grave problem before many more years have passed. As the population of the earth increases, so too does its urgent need for energy and consumer goods. The result is widespread pollution of the atmosphere, surface water and soil, which may produce drastic consequences as a result of climatic change and/or damage to the biosphere. It is vital that research be carried out to establish the extent of this threat, so that an international consensus can be reached on what countermeasures are to be taken. By far the most effective action is based on demographic policy. Industrial measures based on the efficient generation and use

*Board of Directors of the Hüls AG, Marl, Germany

of energy, and action to preserve the earth's vegetation, must be taken at the same time. Alternative energy, as it is known, will play only a minor role, and these priorities must be borne in mind when determining the future focus of research.

Fossil Energy Resources

Our resources will last much longer than the three decades on which I will focus my attention. Table 1 shows the presently known or presumed resources of fossil and nuclear fuels. Table 2 gives ratios of the resources of Table 1 over annual consumption figures (1988). These ranges tell us how long the resources will last, if we continue consumption at the present rate. Increasing consumption would, of course, shorten these ranges, as shown in the last column[1,2,3,4,5]. There will obviously not be a shortage of any of these fuels.

TABLE 1. WORLD ENERGY SUPPLY — 1989
(in billion tons — coal equivalent)

	Deposits	
	Known	Estimated
Mineral oil	182	400
Oil shale, tar sands, heavy crude oil		2,000
Natural gas	146	450
Coal	609	7,500
Nuclear fuel		
130 $/lb U	80	
500 $/lb U		900

TABLE 2. WORLD ENERGY SUPPLY
(estimated ranges in years)

	Known deposits divided by consumption (1988)	Estimated deposits divided by consumption (1988)	Estimated ranges at increasing demand
Mineral oil	40	90	30-50
Natural gas	60	190	150
Coal	170	2,000	1,000

The supply of mineral oil — in its conventional form — will probably decline between 2020 and 2040. But natural gas will last much longer — up to the year 2100 or so — and coal will probably be available for another thousand years.

Heavy crude oils, tar sands, and, above all, oil shales have only been partially explored so far (Canada, Venezuela). They are available in huge quantities, as they probably will be for hundreds of years.

One thing is clear: production costs will be much higher, and oil prices between $25 and $45 per barrel — based on today's prices — are not unlikely ($25 to $30 per barrel in the case of heavy crude oil, including hydrogenation — $35 to $45 per barrel in the case of oil shale, including hydrogenation). Such prices are, however, by no means prohibitive. Or, to put it the other way round: such prices are still sufficiently low to make "alternative energy sources" non-competitive.

As long as the supply of oil exceeds demand, heavy oil reserves will remain untouched. Nevertheless, commercial production and processing technologies are available.

The range of nuclear fuels is somewhere between the ranges for mineral oil and natural gas, if we extrapolate today's consumption. It depends greatly on the efforts we make in the field of uranium production. Using breeder technology, the ranges should become virtually unlimited.

Liquid Fuels

Limited supply of fossil fuels is not to be expected for the next 100 to 150 years. Also, liquid fuels for the transportation sector will not run short.

- Heavy crude oil will be available for a very long time as a raw material for the production of liquid fuels.
- Light mineral oil, presently used in considerable quantities for heating purposes, can easily be replaced by natural gas.
- Methanol can be used as an engine fuel, and it can be made from natural gas or coal.
- Finally, a number of commercial processes are available to make gasoline (petrol) from natural gas, synthesis gas and/or methanol.

Considering all this, developments such as the "electro-car" or the "H_2 Car" cannot be justified on the basis of expected fuel shortages, at least not for a very long time. And it may well be that these developments do not make any sense at all, as it is doubtful whether they lead to a better environmental situation.

Electro-cars do not show superior efficiency with regard to primary energy consumption that causes CO_2 emission. In fact, their total efficiency is probably lower than that of combustion engines, considering the efficiency chain as it is known today:

Electricity from coal: 35%
Distribution of electricity: 90%
High performance battery: 40 to 70%
Electrical drive with control: 60%

Total efficiency is calculated to be in the order of 8 to 13%, thus lower than that of a diesel car. There is certainly no apparent advantage in the electro-car with regard to carbon dioxide emissions as long as electricity from coal is used. And what if electricity were available from other, non-fossil sources?

Well, in that case clean energy with its environmental benefits would be better used by stationary energy consumers (practically without any additional energy losses) than by mobile consumers such as electro-cars, where additional expense and additional energy losses are unavoidable. Therefore: As long as fossil fuels (oil or natural gas) are still being used for domestic heating purposes, clean energy entering the market should first be used in homes, not elsewhere. With respect to clean energy, mobile consumers should only be considered after all stationary energy consumers have switched over to clean fuel.

These remarks apply just as well to hydrogen cars, and, for all these reasons, we conclude that research and development investments in the fields of electro-cars and hydrogen cars do not have priority.

Nuclear energy

If we look into the next century we can see inherently safe nuclear technologies emerging to cover additional energy demand. The high temperature reactor (HTR) concept is one route to achieve these goals. Unfortunately, expenses for its further development are currently blocked by the lack of a political consensus.

There are a number of issues which require our attention: For example, the development of high-temperature resistant fuel spheres which prevent oxidization of the graphite matrix in the case of overheating and simultaneous access of air into the reactor. Another issue: the further development of modular HTR components, which are sufficiently reliable to be suited for export to third world countries.

Regarding nuclear fusion, we are much less optimistic. German experts (K. Pinkau, director of MPI-Plasmaphysik, Garching) estimate that the technology required will be available 50 years from now — at the earliest. One of the most difficult aspects is the complexity and sheer number of peripheral aggregates required, which will most certainly cause a "flight into size." But the torus of the nuclear fusion reactor cannot be enlarged beyond a certain limiting value, for a simple reason: Its wall must allow transfer of the heat of fusion. The bigger the torus, the higher the heat flow per area unit of the torus wall and the higher the temperature load. Thus, a principal barrier is set by the thermal stability of the torus wall. We are not aware of any solution for this problem, but it is conceivable that our view is too pessimistic.

A final word regarding nuclear energy: if we had an abundant and safe source of nuclear energy, we could supply all stationary energy consumers with electricity, satisfying about 70% of the world's total energy requirements.

Renewable Energy Sources

The potential for renewable energy sources will remain relatively small in the period we are considering here — i.e., the next thirty years. It will not exceed 10% of the total energy supply. Eighty percent of the renewable energy supply will be hydropower and energy from waste and biomass combustion, while photovoltaics, wind force and geothermal heat will hardly supply 2% of the total demand in the foreseeable future.

In the case of hydropower, less than half of world potential has so far been put to use. The rapid exploitation of more remote sources is closely coupled with the development of a modern low-loss electricity transport system (HVDC, high-voltage direct current). While the most important technical preconditions have already been met, much research is still needed: improvements in thyristor stations to lower production cost; improvements in switches and cable insulation; development of electrical machines which are directly linked to the HVDC transmission mains (influence machines).

By today's criteria, hydropower is the only renewable energy source which is economically sound, with current prices of less than DM 0.10 per kWh in the case of the larger plants. Increased exploitation of hydropower, for example in Siberia, Africa and Asia, and integrating these resources in an intercontinental HVDC network would allow peak load equilibration across several time zones and could replace a considerable part of today's fossil-based electricity generation.

The technologies of waste and biomass combustion and low-temperature gasification have only now reached a degree of maturity allowing the generation of energy on a break-even basis. Particularly attractive is a new concept of waste gasification, which provides for the autothermal decomposition of a mixture of waste and dry biomass in the presence of air to yield a heating gas (N_2, CO and H_2). After purification, the gas is conducted to the next power station, where it is used for sub-firing. Public acceptance of this technology should be high, as it has no need of gas exhaust stacks. No waste water is generated in this process. However, research is still required, especially with regard to ash disposal.

In densely-populated countries (Germany, Japan, the UK, etc.), a few percent of the total energy requirement can be met by the exploitation of renewable raw materials. Unfortunately, there is little space available for this. The use of biological waste (straw, wood wastes, etc.) and the cultivation of rapidly growing plants, such as elephant grass, appear reasonable, the latter yielding an astonishing 15-30 tons of biomass per hectare. Research issues here are optimizing cultivation and harvesting methods (fertilizing, crop rotation) avoiding soil deteriorations, and so on.

Bio alcohol for Otto engines and rape oil for diesel engines can only acquire regional importance in sparsely populated countries, in view of the low hectare yields involved. Profitability, which has hardly been achieved in Brazil (in spite of considerable government subsidies) is nowadays suffering badly from oil price fluctuations.

Table 3 shows the yields (including energy yields) of various energy plants. Market research and investigation on logistics seem to be the only open questions in this area.

TABLE 3. ENERGY FROM PLANTS — ANNUAL YIELDS

		Net Energy Yield GJ/h	
	Yield t/ha	main product only	waste use included
Rape Seed oil	7.4	16.8	82.8
Alcohol from			
Sugar beet	46.6	28.6	38.6
Wheat	11.4	-4.6	101.4
Potatoes	30.0	17.2	32.2
Whole plant combustion	10.0	134.2	134.2
Combustion of fast growing wood	16.0	155.0	155.0

Excerpt from Energiegutachten Baden-Württemberg, DLR, 1987.

Finally, the overall energy balance of energy generation concepts is a research topic in itself. As far as we can tell today, there is a good chance that there are processes in which the overall energy balance is clearly positive — which would mean that energy is gained without adversely affecting the CO_2 balance. All other types of alternative energy are, for the present, simply much too expensive.

Using photovoltaic processes to generate electricity, for example, is about 10 times as costly as using coal. This is not only due to the low degrees of efficiency of today's photovoltaic cells, but also to the expensive peripheral aggregates needed, such as batteries and transforming equipment. In recent years, the public has been confronted with a number of success stories and future scenarios in this field. But these stories have been rather misleading and created the entirely wrong impression of a genuine and presently available alternative to the technologies presently used. A sober view of the problems, of the technical weaknesses to be eliminated, and of the conceivable fields of application would be more helpful.

Table 4 gives an overview of the problems still to be solved and the general issues in the fields of photovoltaics, solar energy, wind energy and geothermal heat. Of these four alternative energy sources, it is likely that wind and geothermal energy will be the first to reach the threshold of profitability. Unfortunately, however, the potential of these technologies will even then be rather limited, as only few regions are suitable for their application.

TABLE 4. THE PROBLEMS OF ALTERNATIVE ENERGIES

		Problems
Photovoltaics	Solar Farms	Costs of peripheral aggregates, construction and space are high. Decreasing efficiency with increasing temperature. Limited durability.
Solar Heat	Mirror Power Stations	Peripheral aggregates costly. Operation only in direct sunlight. High temperature process, service and maintenance requirement.
Wind Power	Wind Parks	Expensive construction, gear and bearing problems at larger units. Operation irregular, impossible to plan. Only suitable for coastal regions.
Geothermal Energy	Hot Dry Rock Power Stations	Deep drilling (>5000 m) problematic. Even experiments are extremely expensive.

Hydrogen Technology

Hydrogen as an energy carrier is often mentioned in the debate on future energy supplies. In view of its easy storage and transportation, hydrogen is often claimed to be the ideal carrier for the collection of energy found "in the wrong place or at the wrong time". Take solar farms in the Sahara desert as an example.

As an energy carrier, hydrogen would, of course, compete with electricity — and a critical comparison shows that the odds are set against it. The reasons are the following:

- High-voltage DC technology provides for much more profitable long-distance transportation of electricity (with low losses) than is the case with hydrogen pipelines. Table 5 shows a comparison of two transport systems — HVDC transmission versus hydrogen.

- At the same price per calorie, the end-user will prefer electricity. The electricity system has already been installed and paid for. All users are used to electricity and its devices.

- With an expanding electricity network the inclusion of small local electricity generating facilities is becoming less and less difficult and the need for storage facilities (for hours or days) diminishes.

- There will still be a need for seasonal storage of energy, for which hydrogen is totally unsuited, as sufficiently large caverns to hold large quantities of hydrogen do not exist. Seasonal storage is most easily accomplished with coal.

- The liquefaction of hydrogen is much too expensive to be used for storage, 14 kWh being required per Kilogram of liquid hydrogen — that is one third of its heat value.

- Using hydrogen as a fuel in cars or airplanes affords no advantages, neither in the field of the environment, nor in the field of resource protection. It is also not justified economically. An exception may be the use of hydrogen for bus and truck fleets which operate on constant routes in towns resulting in reduction of emissions in the town centers.

- Storage of hydrogen in the form of metal hydrides is too cumbersome for use in motor vehicles, as the storage devices are simply too heavy.

The conclusion is: From both the ecological and the economical points of view hydrogen cannot be a competitive carrier of energy, which means that the various hydrogen research projects currently receiving public subsidies should be critically reviewed. One should keep in mind that there will be no such thing as a "hydrogen world" or even a "solar hydrogen world."

As a raw material for chemistry and as a material used for other technical purposes the importance of hydrogen is still increasing.

TABLE 5. TRANSMISSION OF ELECTRICITY
FROM CENTRAL AFRICA TO EUROPE
Comparison of Costs (Pf/kWh)

	HVDC-System		H_2-System	
	Hydro-Power	Photovoltaic Power	Hydro-Power	Photovoltaic Power
Cost of Electricity Generation	3.0	80	3.0	80
Cost from Transport Losses	0.6	16	2.9	77
Other Transport Costs	3.5	3.5	17	17
Total Cost of Current in Europe	7.1	99.5	22.9	174

Environmental Problems

All in all, the situation is not particularly ominous, seen from the point of view of the availability of resources. The same cannot be said with regard to disposal. World demand for energy is constantly on the increase. This is on the one hand due to increasing population (Figure 6), and on the other hand to the fact that developing nations have a long way to go in order to catch up. They are pursuing a path of industrialization in order to provide food for an ever-increasing population, and it would be presumptuous to deny such needs. Unfortunately, experience shows that the path from primitive economies to High Tech normally passes via a maximum specific energy consumption. The first car is normally an old 'gas-guzzler.' High energy consumption seems to be characteristic of moderately developed countries.

The People's Republic of China plans to increase the consumption of coal by a factor of forty in the period between 1975 and 2075, if nuclear power cannot be used to a significant extent.[6] Adequate supplies are certainly available. Just imagine: China would then be using about five times as much coal as all the world did in 1990. Already today, China is the world's largest producer of coal with a production of more than 1 billion tons per year. Even granting that socialist systems never achieve their economic goals, the perspective is breathtaking in the truest sense of the word. Considering that some other countries, Russia for example, might also increase their consumption of coal, the efforts made in western countries to reduce coal consumption appear to be of little effect by comparison.

Figure 6   World Population Growth

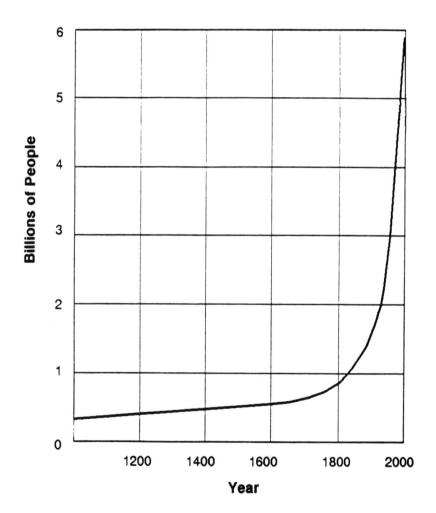

In any event, it will be necessary to use modern technology wherever power stations based on coal are to be constructed. For example, the so-called GuD-power plant systems developed *i.a.* in Germany during the last decade could be used (they are based on a combination of gas and steam turbines, with a high Carnot degree of efficiency, and a total efficiency of the order of 45%). The practical testing of various types of GuD-systems still requires some effort. In addition, the advantages of power/heat coupling (cogeneration, KRAFT/WÄRME-Kopplung) can be used, at least in all cases where industrial complexes are to be supplied with energy. Unfortunately, the practical applicability of cogeneration is frequently overestimated. Economically, uniform heat requirements the year over are an absolute necessity, and this uniformity does not generally exist in residential areas. To eliminate this fundamental disadvantage of power/heat coupling should be a difficult, but rather worthwhile topic for further research (conversion and storage of excess low-temperature energy available in the summer).

As we all know, the generation and use of more and more energy and ongoing industrialization pollute the atmosphere, soil and ground water. The damage done is manifold and it is accumulating. Additionally, the Earth can absorb only a small fraction of this load. Some kinds of emissions can be limited by sophisticated technical means, which are, by nature, expensive.

Think of catalytic converters in modern cars, or of desulfurization and NO_x-reduction measures in power stations, of sewage treatment plants and waste utilization plants with heavy metal concentration. But these technologies are by no means widely used — consistently so only in rich countries: Poverty is dirty.

It is common today to select individual types of emissions and to examine their harmful effects separately: Stratospheric ozone and fluorochlorohydrocarbons, tropospheric ozone and NO_x, pollution in the Mediterranean sea, algal overgrowth, heavy metals in waste material dumps, to mention just a few. However, we must become aware that the total picture is even more complex if the synergies of noxious substances are considered. Not only scientists but also the public should be concerned.

What about the ubiquity of pollution, which is absolutely unavoidable with the world's population growing at the present rate? Research and development work in this sector is certainly needed.

In the eyes of many scientists, the greenhouse effect is the greatest hazard we are facing today. There are, however, dissenting voices. Take recent data from the meteorology institute of the German Max-

Planck-Gesellschaft, which predict only a rather moderate warming of the atmosphere. Who is right? Will the increase be 0.3 degree? Will it be 1.5 degrees, or even 4.5 degrees if the content of greenhouse gases doubles? Maybe we shouldn't rely on computer prognoses alone — in any case they are not understandable to the general public.

Simple physical considerations show that an increase in carbon dioxide must, of necessity, at least cause some increase in temperature. This increase is difficult to quantify — the heat capacity of the continents, and, above all, the oceans has a buffering effect. Motion and exchange of water layers complicate the picture. Increasing evaporation and the formation of clouds will slow the heating process, and the changing albedo (decreases in ice-covered areas, increases in desert areas) will also have an influence. A certain temperature increase seems to have happened in the last century (Figure 7).

Another look in the Earth's past may be helpful — there were extended polar ice caps in the Cambrian Age — even though the CO_2 content of the atmosphere was several percent. The Earth's average temperature was most probably around 20°C — in spite of a tremendous greenhouse effect (Figure 8). We conclude that the latest computer results, which only predict a rather moderate temperature increase, are most likely correct — which does not mean that we are not going to have certain climatic changes.

While the likelihood of catastrophic changes in the Earth's climate is still under debate, another hazard is on the horizon: The biological effects associated with increasing carbon dioxide concentration in the troposphere. Besides the fertilizing action of carbon dioxide (often mentioned as an advantage in American literature), detrimental effects in the biosphere have to be anticipated.

Throughout the Earth's history, the $CO_2:O_2$ ratio had an important and very well-documented influence on the development of the biosphere (Figures 9 and 10). In fact, this ratio was a key parameter of evolution and the high rate of change we are witnessing today suggests that there may be significant consequences. For example: micro-organisms adapt much faster to changing environments than plants do — and the resulting evolutionary asymmetry may be a problem for plant health. We already have a carbon imbalance in our atmosphere, and the imbalance may be further increased by plant diseases and the resulting decrease in the earth's plant cover. There is no doubt that a careful study of global atmospheric changes and their effects on the Earth's biosphere are among the most important issues of research today.

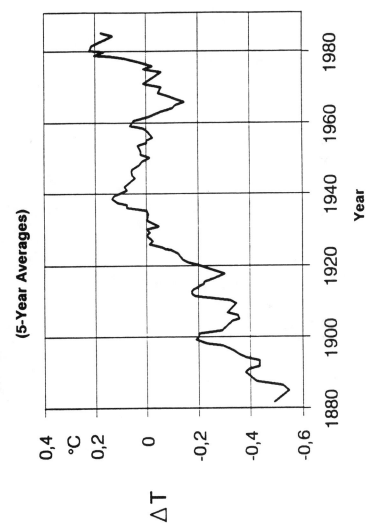

Figure 7   Rise in Global Temperature since 1880

(5-Year Averages)

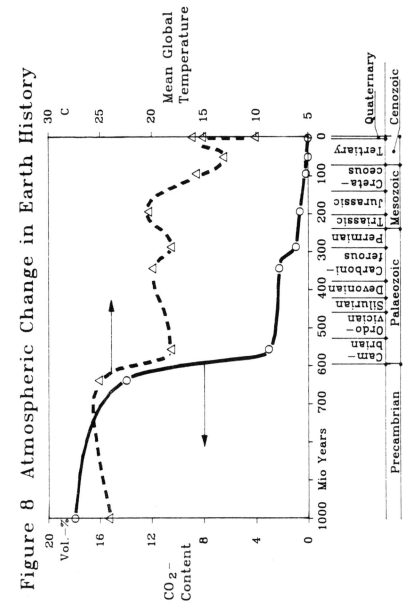

Figure 8 Atmospheric Change in Earth History

Comment on Figure 8

Fossils show that an ice cap developed during the Cambrian Age in the south polar region. The broken line shows that he Earth's average temperature was slightly below 20°C. CO_2 content is given by the continuous line. It was above 2.5% during the Cambrian Age.

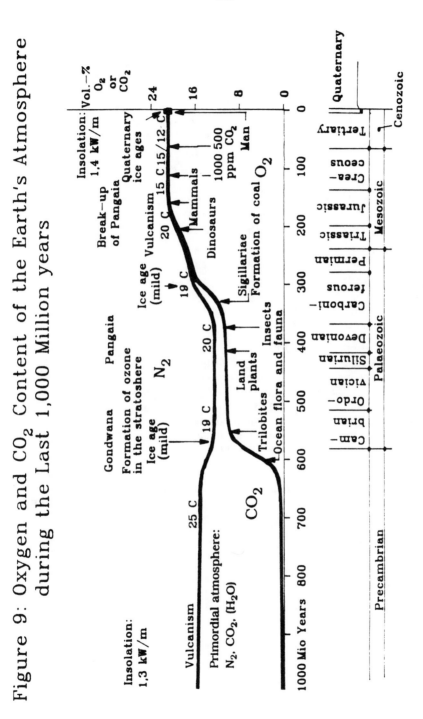

Figure 9: Oxygen and CO₂ Content of the Earth's Atmosphere during the Last 1,000 Million years

Figure 10   Increasing Biological Variety

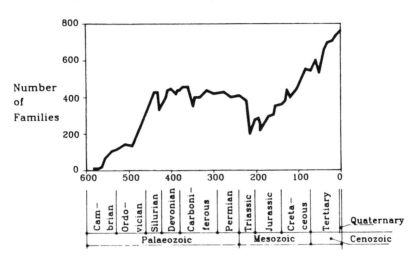

Comment of Figures 9 and 10

The serious atmospheric changes beginning with the onset of the Cambrian Age go hand in hand with the development of biological life forms (Figure 8: Increasing oxygen content, decreasing carbon dioxide content).

The number of animal and plant families increases in two stages (Figure 9), corresponding to two stages of atmospheric change. The development of aquatic life forms occurred during the first stage, and the development of terrestrial life forms during the second stage. Their changing environment, particularly changes in oceanic pH values, forces animal and plant species to adapt. The formation of bones and calcareous shells, for example, is strongly pH-dependent. Only few plant or animal species have survived over periods longer than 2 or 3 million years. Changes in the CO_2 content of the atmosphere beyond 1,500 ppm always necessitated adaptation.

Energy Policy

Carbon dioxide as a topic will gravitate to the center of public attention in the decades to come. This will affect the discussion of energy generation and utilization. Energy sources, which are neutral with regard to the CO_2 balance will be called for, and it is certain that attempts will be made to limit carbon dioxide emissions using both tax incentives and penalties. Obviously, combustion processes and individual traffic will be the focus of criticism. And it is also obvious by now that Germany intends to spearhead the development of anti-CO_2 measures.

While we do see the dangers of increasing CO_2 levels, we would suggest to proceed with caution. If Germany wishes to give new impulses to the rest of the world, it should better be the right ones. Unfortunately, we can see little consensus among experts, so far. I would like to point out some of the issues, as I see them:

• Almost all biological processes are in some way "CO_2-relevant". Therefore, there will be a network-like interdependence of CO_2 effects. It is difficult to view processes in isolation, and it would be wrong to narrow the problem down to coal firing in power plants and transportation.

• Emissions and remissions of CO_2 are equally important for the atmospheric CO_2 balance — remissions being the conversion of CO_2 to plant mass, i.e., carbohydrates by means of photosynthesis.

It would be foolish to ignore remission processes — this would jeopardize success, as there is a strong interdependence. Globally, the decline of remissions of CO_2 is more dramatic than the increase of emissions. The effect that a country has with regard to the world's CO_2 balance does not depend only on its emission (or per capita emission) — it depends on the ratio of emission to remission.

• Industrial measures alone cannot solve the CO_2 problem if the world population continues to grow at its present rate and the "consumption backlog" of the developing countries is to be abolished.

The effect which a country has on the CO_2 balance depends not necessarily on the living standard of its population. The ratio of population:number of trees is much more important. Countries which require heating in the winter show relatively high CO_2 emissions. Countries without forests have minimal remission values. In both cases the influence on the CO_2 balance is negative.

Figure 11 shows only the emissions — if we had included both emissions and remissions, the USSR would get a much more positive rating, and Japan a much more negative one.

Figure 11 CO$_2$ Emissions (per capita) and Gross National Products for Various Countries and Regions – 1985

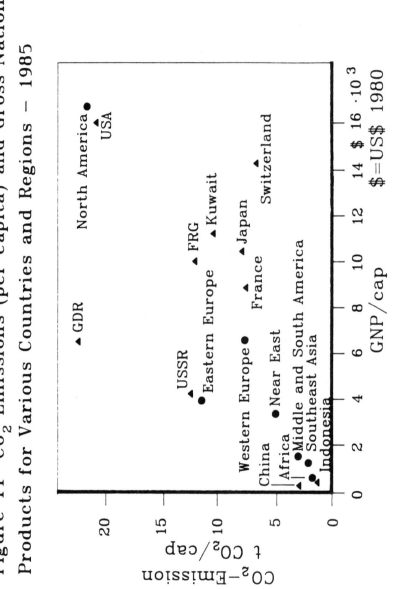

- It is a difficult task to single out processes and optimize them with regard to CO_2 emissions. Saving energy has to be the major goal. Normally, saving energy and lowering CO_2 emissions go hand in hand. There is also one caveat — be careful with high technical investment to limit CO_2 emissions — technical investments have their own effect on the CO_2 balance.

- Both energy savings and reforestation are steps in the right direction. For example, 80 to 90% of the energy used today for the heating of buildings could be saved using the well-known methods of insulation and passive solar heating. Obviously, this would require a revolution in the field of architecture.

- To capture CO_2 from combustion processes, sending it to the bottom of the ocean in the form of dry ice, a suggestion which has recently been published, is, in our opinion, utopian.

- The most reliable way to measure progress with regard to the CO_2 balance is to compare national primary energy consumptions. The Federal Republic of Germany has shown a decreasing consumption during the last 20 years — even without a carbon dioxide tax.

- So far, there are widely differing views regarding the effects of the increase in atmospheric CO_2. It would be helpful to clarify controversial issues by means of research programs and to achieve some sort of consensus — because, with a consensus on the issues, significant counter-measures will be accepted much more readily.

- I beg your pardon but it will not be possible to be successful without also addressing the problem of over-population.

Conclusion

The question we asked in the beginning was:

> What kind of energy policy will minimize global
> environmental damage in the next century?

There is a clear answer: The problem of global environmental damage certainly cannot be solved by the industrialized countries alone nor can it be solved by technical means alone. Technical "actionism" may even be harmful, because it would generate an ill-founded feeling of security in the public — such as "technology will take care of it." What is urgently required is the development of a consensus in the scientific and political communities regarding the environmental dangers to be expected. A significant amount of basic research is required to further clarify the issues.

Diagnosis should come before therapy, and the patient should believe in the diagnosis. It will not be possible to take measures on a global scale before consensus has been achieved on the issues. Obviously, the prime issue is stopping the growth of the world population.

This is a tremendous sociological task connected with immense psychological barriers. Finding effective actions is of course a subject of research. The most important measure on the technical side will be supporting developing countries: to avoid further deforestation and to use environmentally acceptable technology in their further development. Know-how transfer in the field of environmental protection will require considerable effort on the part of the industrialized nations.

The accompanying measures in our own country are obvious: as far as possible good housekeeping with energy, improving the state of health of our forests, and reducing the emissions in energy production, specifically the CO_2 emission.

Although presently unpopular, we have to do research in the field of nuclear energy in order to develop even safer reactor types. Industrial processes should be developed which are more selective and less energy-intensive than some of the processes used today and which are suitable for use in the third world. This involves, among others, the implementation of sophisticated chemistry.

Literature

1. Daten und Fakten zur Energieversorgung, Bayerisches Staatsministerium für Wirtschaft und Verkehr, 12, 1989.

2. Hafele, W., Energiesysteme im Übergang, MI Poller, 1990.

3. Oeldorado, ESSO AG, Hamburg, 1989.

4. Zahlen aus der Mineralölwirtschaft Deutsche BP AG, Hamburg, 1989.

5. Heinloth, K., *Physik unserer Zeit, 18*, 2, 47-51, 1987.

6. Lu Yingzhong, Institute of Nuclear Energy Technology, Beijing, China, Lecture given on December 14, 1988 in Aachen (Institute of Technology), Germany.

7. Markl, H., Energiewirtsch. *Tagesfragen, 38*, 8, 582, 1988.

8. Hansen, J., and Lebedeff, S., *Geophys. Res. Lett., 92*, D11, 1334, 1987.

9. Beckmann, G., B. Klopries, *Agrarische Rundschau, 1*, March, 32, 1989.

10. Wagner, H.J., CO_2-Emissionen und Energieversorgung, Sonderdruck Forschungszentr. Jülich GmbH, 1990.

INTERNATIONAL COOPERATION AND SOME RESEARCH NEEDS TO IMPROVE OUR UNDERSTANDING OF THE CHEMISTRY OF THE ATMOSPHERE

Valentin A. Koptyug*

As is known, CHEMRAWN is the acronym for **Chem**ical **R**esearch **A**pplied to **W**orld **N**eeds. The purposes of IUPAC CHEMRAWN Conferences are identifying and supporting basic and applied research that help to solve global human problems.

Among many global needs which the humankind has recognized by the end of the 20th century, the most important one is the need for sustainable development in a situation when the world population is growing, natural non-renewable resources are becoming more and more limited, and environmental degradation is becoming more and more visible and dangerous. The cumulative impact of humankind's activities has reached a point where the life on Planet Earth is at risk from global environmental changes.

This is the reason why the United Nations decided to convene in June 1992 a special UN Conference on Environment and Development (UNCED) in order to analyze the international collaboration concerning global changes and to work out a program of actions for the 21st century based on the scientific approach (AGENDA-21).

Among other important topics, the UNCED will pay attention to the protection of the atmosphere with emphasis on the greenhouse effect, ozone layer depletion and transboundary air pollution.

*Professor, Presidium of Academy of Sciences, Moscow, Russia

Therefore, the CHEMRAWN VII Conference "The Chemistry of the Atmosphere: Its Impact on Global Change" may be considered as a part of the preparatory work of the world chemical community for the UN Conference.

In preparatory work for UNCED, the International Union of Pure and Applied Chemistry has also launched a broader mission-oriented Program on Chemistry and Environment.[1] We are now studying possible liaisons of this program with other international programs and trying to recognize the areas which are not sufficiently covered by the existing programs relating to the chemical aspects of environmental pollution.

Environmental problems, including atmospheric pollution, have been attracting attention of the United Nations since the 1970s. Their anxiety on the state of the atmosphere is illustrated in particular by the documents listed in Table 1.

TABLE 1. INTERNATIONAL AGREEMENT IN THE FIELD OF ATMOSPHERE PROTECTION[2]

Convention on Long-Range Transboundary Air Pollution, Geneva, 1979.

- Protocol to the 1979 Geneva Convention. Long-Term Financing of the Cooperative Program for Monitoring and Evaluation of the Long-Range Transmission of Air Pollutants in Europe (EMEP), Geneva, 1984.

- Protocol to the 1979 Geneva Convention. Reduction of Sulfur Emissions or their Transboundary Fluxes by at least 30 per cent, Helsinki, 1985.

- Protocol to the 1979 Geneva Convention. Control of Emission of Nitrogen Oxides or their Transboundary Fluxes, Sofia, 1988*.

Vienna Convention for Protection of the Ozone Layer, Vienna, 1985.

- Montreal Protocol on Substances that Deplete the Ozone Layer, Montreal, 1987.

- Amendment to the Montreal Protocol on Substances that Deplete the Ozone Layer, London, 1990.

*There is also a 1991 draft protocol on volatile organic compounds.

The current UN activity in this area is concentrated on three main aspects of atmospheric pollution:

- estimating the effects of increasing concentrations of greenhouse gases on the Earth's climate and recognizing ways to avoid possible negative consequences;
- supporting monitoring and developing recommendations on how to keep an "ozone umbrella" against dangerous ultraviolet radiation;
- developing recommendations and reaching an agreement on how to prevent acid rains.

The UNCED Preparatory Committee has presented at its 3rd session (August 1991) a set of documents with the analysis of:

- interrelations between air pollution, ozone layer depletion and climate changes;[3]
- existing systems of monitoring transboundary air pollution and a basis for international actions;[4]
- recent scientific findings on ozone depletion and a basis for international actions;[5]
- probable consequences of climate change;[3,6]
- possible ways of restructuring systems of energy production and use, and of transportation (as the key sources of atmospheric problems and climate change) in accordance with the requirements of sustainable development;[3]
- most important goals that rise from the preceding item for the AGENDA-21 (Agenda of Science for Environment and Development in the 21st Century).[7]

The above-mentioned global aspects of atmospheric pollution are fundamental for many existing international programs[8], including the International Global Atmospheric Chemistry (IGAC) Program.[8,9] This program is a core project of the well known International Geosphere Biosphere Program (IGBP) established by ICSU in 1985 and seeks to understand quantitatively the chemical and physical processes that determine the atmospheric composition. The structure of the IGAC Program is shown in Table 2. It is mainly based on the earlier created program of the International Association of Meteorology and Atmospheric Physics and partly overlaps with the SCOPE project on Trace Gas Exchange between the Biosphere and Atmosphere.

TABLE 2. STRUCTURE OF THE IGAC PROGRAM

I. Natural Variability and Anthropogenic Perturbations of the Marine Atmosphere:

- North Atlantic Regional Study,
- Marine Aerosol and Gas Exchange-Interaction with Atmospheric Chemistry and Climate,
- East-Asian-North Pacific Regional Study.

II. Natural Variability and Anthropogenic Perturbations of Tropical Atmospheric Chemistry:

- Biosphere-Atmosphere Trace Gas Exchange in the Tropics,
- Deposition of Biogeochemically Important Trace Species,
- Impact of Tropical Biomass Burning on Atmospheric Chemistry and Biogeochemical Cycles,
- Chemical Transformations in the Tropical Atmosphere and Their Interaction with the Biosphere,
- Rice Cultivation and Release of CH_4 and N_2O.

III. The Role of Polar Regions in Changing Atmosphere Composition:

- Polar Atmospheric Chemistry,
- Polar Air-Snow Experiment.

IV. The Role of Boreal Regions in Biosphere-Atmosphere Interaction:

- High-Latitude Ecosystems as Sources and Sinks of Trace Gases and their Sensitivity to Environmental Distribution.

V. Trace Gas Fluxes in Mid-Latitude Ecosystems.

VI. Global Distributions, Transformations, Trends and Modeling.

- Global Tropospheric Ozone Network,
- Global Atmospheric Chemistry Survey,
- Development of Global Emission Inventories.

VII. Cloud Condensation Nuclei as Controllers of Cloud Properties.

VIII. Intercalibration/Intercomparison.

In many cases the IGAC Program will be built on the existing national programs. This explains a regional character of the main part of the projects.

Very close to the IGAC Program is the Stratosphere-Troposphere Interactions and Biosphere (STIB) Program.[8] Its structure is shown in Table 3.

TABLE 3. STRUCTURE OF THE STIB PROGRAM

I.   Stratospheric Changes and the Penetration of UV-Radiation

II.   Stratosphere-Troposphere Exchange

III. Anthropogenic Trends and Natural Variability

IV.   Stratospheric Aerosols and Their Climate Effects

V.   The Impact of Stratospheric Changes on Climate

If we now look at the agenda of the CHEMRAWN VII Conference, we will see that it corresponds to the well recognized global problems of atmospheric pollution which are considered through the prism of chemical science. Chemistry as a scientific branch of knowledge is concentrated in this case on photochemically induced free radical reactions in which quite simple species such as molecules of oxygen, ozone, nitrogen, sulfur oxides, and chlorofluoromocarbons are involved. Many of corresponding elementary reactions are or can be well studied, but their combination in the open dynamic system of the atmosphere leads to many difficulties and uncertainties in the estimation of global results.

I will not further discuss the main problems of global changes and their causes. Instead I would like to draw attention to other important aspects of chemical pollution of the atmosphere which in my opinion introduce some additional complications and are not adequately covered by international programs and projects.

The atmosphere of the Earth is a life-supporting and life-protecting system. But in our days it has also turned into a reservoir for many by-products of human activity. This reservoir is simultaneously playing the role of a flow reactor for chemical transformation of pollutants and the role of a transport system for delivering pollutants and products of their transformation, many of which are harmful, to sensitive parts of the ecosystems.

Therefore, we should take into consideration not only a change of protecting properties of the atmosphere ("greenhouse" effect, destroying "ozone umbrella") but also the distribution of various harmful substances and products of their transformation through atmospheric channels. In this area the greater part of international efforts are directed to the problem of acid rains and its origins.

But we should not forget other dangerous groups of pollutants. Among them are metals.

There exists a danger of poisoning the biosphere by global metal pollution.[10,11] This type of pollution involves not only soil and aquatic systems but also the atmosphere. Tables 4 and 5 give a general picture of trace metals emissions from natural and anthropogenic sources to the atmosphere. As is seen from these data, the anthropogenic emissions have become dominant for most trace elements in the atmosphere. Anthropogenic emissions of lead, cadmium, vanadium and zinc exceed the fluxes from natural sources by 28-, 5-, 3- and 3-fold, respectively. Industrial contribution of arsenic, copper, mercury, nickel and antimony are 100 to 200 percent of the emissions from natural sources.

Of course, there are some uncertainties in the estimation of global emissions of heavy metals to the atmosphere (cf. Table 6) due to large amounts of natural and anthropogenic sources with different, and, in some cases, variable levels of emission, but general conclusions are reliable.

The International Institute for Applied Systems Analysis has emphasized: "As with acid pollutants, atmospheric emissions of heavy metals are an international problem, often travelling 1,000 or 1,500 kilometers before deposition."[12]

The problem is really international, but at the same time a coordinated, international heavy metals monitoring program is not yet launched, as can be judged from the data of Table 7, which summarized the number and location of the WMO GAW stations.[4]

The global WMO Background Air Pollution Monitoring Network (BAPMoN) run by WMO and UNEP has traced long-range transboundary air pollution since 1968 and has provided most of the atmospheric data to the UNEP Global Environment Monitoring System (GEMS). Together with the Global Ozone Observing System, it is now a part of the WMO Global Atmosphere Watch (GAW), which operates a network of 337 stations in 78 countries.[4]

The well developed Cooperative Program for Monitoring and Evaluation of the Long-range Transmission of Air Pollutants in Europe (EMEP), in its fifth-phase projects (1990-92), includes gas and particles measurements of SO_2, $SO_4^=$, NO_2, O_3, NO_3^-, NH_3, NH_4^+, NHO_3 and precipitation measurements of pH, $SO_4^=$, NO_3, Cl^-, NH_4^+, K, Na, Mg, and Ca. But the measurements of heavy metals are still in the stage of planning.

TABLE 4. WORLDWIDE EMISSIONS OF TRACE METALS FROM NATURAL SOURCES TO THE ATMOSPHERE
(thousand tonnes per year)[10]

Elements	Wind-borne soil particles	Sea salt spray	Volcanoes	Forest fires	Biogenic sources	Total
Antimony	0.78	0.56	0.71	0.22	0.29	2.6
Arsenic	2.6	1.7	3.8	0.19	3.9	12
Cadmium	0.21	0.06	0.82	0.11	0.24	1.4
Chromium	27	0.07	15	0.09	1.1	43
Cobalt	4.1	0.07	0.96	0.31	0.66	6.1
Copper	8.0	3.6	9.4	3.8	3.3	28
Lead	3.9	1.4	3.3	1.9	1.7	12
Manganese	221	0.86	42	23	30	317
Mercury	0.05	0.02	1.0	0.02	1.4	2.5
Molybdenum	1.3	0.22	0.40	0.57	0.54	3.0
Nickel	11	1.3	14	2.3	0.73	29
Selenium	0.18	0.55	0.95	0.26	8.4	10
Vanadium	16	3.1	5.6	1.8	1.2	28
Zinc	19	0.44	9.6	7.6	8.1	45

TABLE 5. WORLDWIDE EMISSIONS OF TRACE METALS FROM ANTHROPOGENIC SOURCES TO THE ATMOSPHERE
(thousand tonnes per year)[10]

Elements	Mining + smelting & refining	Manufacturing processes + commercial uses*	Energy production	Waste incineration	Total
Antimony	0.10+1.42	...	1.30	0.67	3.5
Arsenic	0.06+12.3	1.95+2.02	2.22	0.31	19
Cadmium	...+5.43	0.60+...	0.79	0.75	7.6
Chromium	...	17.0+...	12.7	0.84	31
Copper	0.42+23.2	2.01+...	8.04	1.58	35
Lead	2.55+46.5	15.7+4.50	12.7**	2.37	332
Manganese	0.62+2.55	14.7+...	12.1	8.26	38
Mercury	...+0.13	...	2.26	1.16	3.6
Nickel	0.80+3.99	4.47+...	42.0	0.35	52
Selenium	0.16+2.18	...	3.85	0.11	6.3
Thallium	...	4.01+...	1.13	...	5.1
Tin	...+1.06	...	3.27	0.81	5.1
Vanadium	...+0.06	0.74+...	84.0	1.15	86
Zinc	0.46+72.0	33.4+3.25	16.8	5.90	132

*Including agricultural use.

**Plus 248 thousand tonnes for transportation

TABLE 6. COMPARISON OF TOTAL EMISSIONS OF TRACE METALS IN THE ATMOSPHERE ACCORDING TO TWO RECENT REVIEWS
(thousand tonnes per year)

Element	Sources			
	natural*	anthropogenic*	natural**	anthropogenic**
Antimony	2.6	3.5	1 (0.5-1.8)	24 (18-38)
Arsenic	12	19	8 (3-13)	40 (25-80)
Cadmium	1.4	7.6	1 (0.3-7)	7.7 (5.5-11)
Chromium	43	31	60 (44-130)	50 (21-94)
Copper	28	35	20 (18-22)	140 (56-260)
Lead	12	332	27 (4-45)	425 (300-470)
Manganese	317	38	600 (516-750)	215 (107-320)
Mercury	2.5	3.6	20 (2.5-150)	6 (1.7-11)
Nickel	29	52	27 (8.5-54)	80 (43-98)
Selenium	10	6.3	10 (6-14)	7 (1.1-11.7)
Vanadium	28	86	65 (40-79)	170 (110-210)
Zinc	45	132	90 (36-200)	500 (315-840)

*Nriagu, 1990.[10]

**Malachov and Machonko, 1990.[11] (Values in parentheses indicate a spread of data of various authors).

TABLE 7. NUMBERS OF GAW STATIONS[4]
(by 31 December 1990)

Parameters measured	Africa	Asia	South America	North & Central America	Pacific (SW)	Europe	Antarctic	All regions
Precipitation chemistry	14	17	7	41	14	74	1	168
Particles	8	5	2	11	20	33	1	80
Sulfur dioxide	1	1	1	-	2	28	-	34
Oxides of nitrogen	-	1	-	1	4	20	9	28
Carbon dioxide	4	1	1	14	6	12	2	40
Ozone	9	43	5	17	11	49	7	141
Heavy metals	-	-	-	-	-	4	-	4

Regions

Efforts are now underway to develop global inventories under an atmospheric chemistry project within the International Geosphere-Biosphere Program of ICSU.[4,8]

The worldwide contamination of the environment (air, water and soils) with toxic metals (especially with Pb, Cd, Hg, and As) is a matter of concern. In many urban areas and around some point sources the natural emissions are insignificant in comparison with the anthropogenic metal pollution. The influence of heavy metals on human health is usually considered from the point of view of acute rather than chronic effects. In the present situation, the long-term effects of exposing human populations to small doses of toxic metals in the environment should receive adequate attention. The reliable information for the estimation of postponed effects of small doses of toxic elements can be obtained through medical surveys of population (including genetic alterations) in regions of geochemical anomalies, characterized by high contents of heavy metals, with the simultaneous investigation of surrounding ecosystems, taking into account the tendency of some heavy metals to accumulate in components of ecosystems. Therefore, it is believed that launching joint projects of this type by the WHO, UNEP ("Biogeochemical Cycles" and "Health and Toxicology" programs) and IUPAC will be very important for understanding the scale of hazard.

The main sources of heavy metals pollution of the biosphere, including the atmosphere, are connected with mining, smelting and refining in non-ferrous metallurgy, and with energy production and the transport sector (exhaust gases of cars and trucks, burning leaded gasoline).

The most important way to preclude anthropogenic changes of the atmosphere is the reduction of emissions of harmful chemical substances by industry, power generation, transport, and municipal utilities. Introducing some international restrictions on emissions is demonstrated by the above-mentioned conventions and protocols. Of course this should be supported by vast technological changes in the area of production. Therefore, a giant problem emerges — how to develop industry in the context of a new vision of the future in connection with environmental problems and in the context of sustainable development of our civilization.

This is the reason why the next CHEMRAWN VIII Conference will be devoted to Chemistry and Sustainable Development (sub-heading — Towards a Clean Environment, Zero Waste and Highest Energy Efficiency).

The toxic effects of heavy metals is one side of the problem of global environmental pollution, including atmospheric, by trace metals. Another side — a catalytic effect of many metals — has direct relation to the chemistry of the atmosphere. The role of catalysis in the atmospheric chemistry is an area that is open for fruitful international scientific collaboration. The same is related to the more general area — the investigation of the role of aerosols and heterogeneous processes in the chemistry of the atmosphere[13] (N 11, p. 1729).

Another important direction of international activity seems to be the creation of kinetic data bases for adsorption processes, gas-phase and surface reactions, and photochemical transformations. The IUPAC Commission on Chemical Kinetics (former Chairperson Prof. E. T. Denisov, now Dr. J. T. Herron) is trying to join efforts of specialists of many countries in this direction. In April 1991, the special international workshop "Databases in Chemical Kinetics" was held in Novosibirsk and the situation in this area was discussed.

Databases on chemical kinetics are very important for developing mathematical models of chemical processes in the atmosphere[13] (N 10, p. 1627; N 11, p. 1757).

It is my pleasure as the editor-in-chief of the Russian review journal "Uspekhi Chimii" (Advances in Chemistry) to inform you that after we had decided to devote two issues of this journal to the chemistry of the atmosphere, a group of specialists invited by Prof. Yu. N. Molin, covered, among others, almost all of the above-mentioned "hot" areas by prepared papers.[13]

Now I would like to return to the IUPAC Chemistry and the Environment Program and to pose a question — what is being done and what can be done under the aegis of the IUPAC in relation to the chemistry of the atmosphere?

First of all, it is necessary to stress the importance of this CHEM-RAWN VII Conference that should additionally stimulate international collaboration in many of the discussed areas. The recommendations that will be developed by the Future Action Committee will serve as a guide in this joint work.

A list of ongoing IUPAC projects relating to the problem under discussion is given in Table 8.

TABLE 8. IUPAC PROJECTS RELATING TO THE
CHEMISTRY OF THE ATMOSPHERE

I. **Analysis of Situation and Methodological Aspects**

- Compendium of Agencies, Institutes and Ongoing Activities in the Field of Atmospheric Chemistry
- Inventory of Regulations for Emissions and Standards on Ambient and Workplace Atmosphere
- Inventory of Current Tropospheric Sampling Programs PAC, 62, N 1, 163-176 (1990)
- Evaluation and Harmonization of Current Tropospheric Sampling Networks Worldwide
- Assessment of Uncertainties in the Projected Concentration of the Carbon Dioxide in the Atmosphere, PAC, 63, N 5, 764-796 (1991)
- Inventory of Missing Emission Data Necessary to Evaluate Global Atmospheric Changes
- The Use of Passive Samplers for Monitoring Atmospheric Constituents
- Major Concerns and Research Needs for our Understanding of the Chemistry of the Atmosphere

II. **Nomenclature and Units**

- Glossary of Atmospheric Chemistry Terms, PAC, 62, N 11, 2167-2219 (1990)
- Glossary of Terms Used in Environmental Analytical Chemistry
- Evaluation and Recommendation of Units for Use in Atmospheric Chemistry

III. **Kinetic Data**

- Evaluated Kinetic and Photochemical Data for Atmospheric Chemistry, *J. Phys. Chem. Reference Data*, 18, 881-1087 (1989)
- Kinetic Data Evaluation for Application in Modeling Studies of Global Atmospheric Chemistry

IV. **Characterization, Transport and Reactions of Atmospheric Particles**

- Source Apportionment of Atmospheric Particles

- Sampling of Airborne Particulate Matter for Analysis
- Sampling and Analysis of Carbonaceous Particles in the Atmosphere
- Characterization of Individual Environmental Particles by Beam Techniques
- Characterization of Environmental Particle Surface by Fourier-Transform Infrared and Nuclear Magnetic Resonance Spectroscopies
- Characterization of Particle Surface Charge
- Microanalysis of Individual Aerosol Particles
- Characterization of the Cr(III)/Cr(VI) Ratios in Aerosols
- Interaction of Electromagnetic Radiation with Airborne Particulate Matter
- Acid-Base Equilibria on Particles in the Atmosphere
- Mass Transport by Airborne Particulate Matter

V. **Other areas**
- Recommendation for the Determination of pH in Acid Rain
- Analysis of Wet Deposition (Acid Rain): Determination of the Major Anionic Constituents by Ion Chromatography, PAC, 63, 907-915 (1991)
- Pesticides in Air

The projects of Section I and partly Section II are coordinated by the Commission on Atmospheric Chemistry (the former Chairman, Dr. J. G. Calvert, now Dr. J. Slanina), of Section III — by the Sub-Committee on Gas Kinetic Data Evaluation for Atmospheric Chemistry (Chairman, Prof. J. A. Kerr) and of Section IV — by the Commission on Environmental Analytical Chemistry (Chairman, Dr. J. Buffle).

I would like to draw attention mainly to Sections I and IV.

The workshop organized in July 1990 by the Commission on Atmospheric Chemistry on the assessment of uncertainties in the projected concentrations of carbon dioxide in the atmosphere demonstrated that the predictions of trends in increasing radiatively active gases and their greenhouse effect are subject to a large margin of uncertainty[14] (cf. 15). The participants of this workshop identified the most important causes for the uncertainty in the projection of future carbon dioxide concentrations and proposed corresponding recommendations for future research.

The methodological consistency of potential global change assessment is very important, taking into account the possible scale of the economic and social response to be required. Therefore, the scientific community should pay adequate attention to the analysis of all basic data and models used for forecasting.

The importance of this area for the international collaboration may be additionally demonstrated by the existing disagreement in the relative roles of the effect of anthropogenic chemical substances and natural processes in "ozone hole" formation.

The projects of Section IV illustrate the IUPAC entry into the area of heterogeneous chemistry of the atmosphere. The ongoing projects are mainly related to the problem of characterization of particles and their surfaces. The reactions on the surfaces, the catalytic role of components, including heavy metals, and the fate of products are waiting for the initiatives of the chemical community in cooperative efforts.

It is also desirable to mention the importance of a study of the role of aerosols in the transport and transformation of radioactive substances (SCOPE project on Biogeochemical Pathways of Artificial Radionuclides Migration) and of using the chemical composition features of particulate matter, emitted by industry, especially particles containing heavy metals, as tracers for the identification of sources of emission at long distances.

A proposal of the International Union of Testing and Research Laboratories for Materials and Structures (RILEM) to the IUPAC to organize collaboration on the problems of atmospheric actions on various materials, including the materials for the facades of buildings, should be taken into account as well. This is also a part of heterogeneous atmospheric chemistry.

In conclusion, I would like to stress once more that the transition to sustainable development requires a significant strengthening of the interaction between chemical science and industry in all areas, and in connection with this to attract your attention to George Whitesides' papers "What Will Chemistry Do in the Next Twenty Years?" and of Wolfgang Jentzsch "What Does Chemical Industry Expect from Physical and Industrial Chemistry" presented to the Jubilee Symposium devoted to the BASF 125 year anniversary.

References

1. Chemistry and the Environment. The IUPAC Program, 1990.
2. Register of International Treaties and other Agreements in the Field of the Environment, UNEP, Nairobi, May 1991.

3. Protection of the Atmosphere Sectoral Issues. Documents of the third session of the UNCED Preparatory Committee, A/CONF.151/PC/60, June 1991.

4. Protection of the Atmosphere: Transboundary Air Pollution, ibid, A/CONF.151/PC/59

5. Protection of the Atmosphere: Ozone Depletion, ibid, A/CONF.151/PC/58.

6. Protection of the Atmosphere: Climate Change, ibid, A/CONF.151/PC/57.

7. Protection of the Atmosphere: Options for AGENDA-21, ibid, A/CONF.151/PC/42 Add 1.

8. Compilative Information on International Organizations and Programs Relating to Chemical and Biochemical Aspects of Environmental Problems. IUPAC and Siberian Branch of the USSR Academy of Sciences, 1991.

9. Chemistry and the Environment: Proceedings of Regional Symposium, Brisbane, 1989. B.N. Noller and M.S. Chadha (eds.), Commonwealth Science Council, UK, 1990.

10. Nriagu, J.O., Global Metal Pollution, *Environment*, 32, 7-11, 2-33, 1990.

11. Malachov, S.G., and E.P. Machonko, *Uspekhi chimii (Russian journal)*, 59, 1777-1798, 1990.

12. Alcamo, J., IIASA Option, September 1991, p.13.

13. *Uspekhi chimii (Russian journal)*, 59, N9, N10, 1990.

14. Slanina S., *Chemistry International*, 13, N1, 10, 1991.

15. Kondratjev, K.Ya. *Uspekhi chimii (Russian journal)*, 59, N10, 1587, 1990.

16. Whitesides, G., *Angew. Chem. Int. Ed. Engl.*, 29, 122, 1990.

17. Jentzsch, W., ibid, 1209.

Global Change
and the
Role of Governments

Sir Crispin Tickell*

The twin problems of conferences of this kind are:

- How to formulate conclusions and recommendations in a way that people, businesses and governments can understand and act upon;
- How to convey such conclusions and recommendations to the policy-makers and decision-takers.

Scientists are not necessarily good at simplifying their thoughts, using plain language, abandoning shades and qualifications, or indeed committing themselves one way or another. Politicians and business leaders and others have their problems, too.

- They like clear and unambiguous advice;
- They operate on short time scales;
- Few of them understand science and its methodology: their mental structures are different;
- They have a lot of other things to think about.

The bridge between long-term science and short-term policies can be long and fragile. In some countries it scarcely exists at all. Scientific advisers do their best, but the cogs of the wheels do not always mesh.

The problem is still greater at the international level, which is usually one further move from reality:

---

*Warden, Green College, Oxford, UK

- Different governments see things very differently;
- There is less of a common sense of interest;
- Agreement on common action is always difficult; no one wants to be put at a disadvantage by acting alone, particularly in environmental matters when pollution can all too easily pay.

So when scientists, people, governments and the international community are faced with the complexities of global change, their first need is for a common base, some fundamental ideas on which all can broadly agree. I will not pretend such a base yet exists. Yet we have something like one, at least in the international sphere, in the form of the report of the World Commission on Environment and Development, chaired by Gro Brundtland, Prime Minister of Norway, and published on 20 March 1987.

The central idea of the Brundtland report was "sustainable development," defined as "development that meets the needs of the present without compromising the ability of future generations to meet their own needs;" and development itself — a very elastic term — was described as involving "a progressive transformation of economy and society" for "the satisfaction of human needs and aspirations."

Of course these definitions raise more questions than give answers, but it would be unwise to dismiss them as word chopping or meaningless rhetoric. From them some key concepts emerged. Here are some I culled from a recent re-reading of the Report:

- the notion of intergenerational equity;
- the need to avoid endangering "the natural systems that support life on earth;" the atmosphere, the waters, the soils and the living beings;
- the ultimate limits to growth and the need to change its quality;
- the notion of a capital stock of resources, (some of them non-renewable) which must be conserved and enhanced;
- global interdependence, and the need to define and promote the common interest;
- the need to ensure a stable level of population;
- the need to reorient technology and manage risk; and
- the need to merge "environment and economics in decision-making" at both the national and international level.

Let us stand back and look at the nature of the problems comprised in environment and development. Here are some simple points to give a sense of perspective:

- In the last 2½ million years, the earth has been in an ice age mode;
- The last 10,000 years have seen all human civilization;
- In the last 250 years the industrial revolution has changed the face of the planet. It is based on an unprecedented consumption of natural resources, especially fossil fuel;
- The last 20 years have seen growing awareness of some of the consequences. This awareness was of course most eloquently expressed by the Brundtland Commission, and has since become widespread, even among those reluctant to admit the consequences.

What are the consequences? In the countries which pioneered the industrial revolution, there has been an amazing rise in living standards which the rest of the world now wishes to emulate. Economic wealth on a familiar definition rose at an almost incredible rate during most of this century.

The success of the industrial countries was founded on their ability to feed their growing populations. They each had an agricultural revolution before an industrial one. Others have not done so well. Total world population rose from 2 billion in 1930 to 5.3 billion now, and will rise again to over 8 billion in 2025. But the ability to feed this population is in doubt (more and more poor countries, for example, in Africa and Latin America, have to import food). The prospect of any substantial rise in living standards in countries without the resources and skills of industry must be illusory.

The carrying capacity of the earth is inevitably a relative if not subjective concept. But some recent calculations are of interest, and I quote from Norman Myers:

- If we all had a vegetarian diet and shared our food equally, the biosphere could comfortably support around 6 billion people;
- If 15% of our calories came from animal products (and again food were shared equally), the figure would come down to 4 billion people;
- If 25% of our calories came from animal products, then it would fall to 3 billion;
- And if 35% of our calories came from animal products, as in North America today, then it would fall to 2.5 billion.

So even if all sorts of improvements could be envisaged the prospect of a rise in human population to 6, 7, 8 billion and upwards is alarming indeed. Such a rise is more than a prospect. Even allowing for war, famine, and disease, the rate of increase — at present some 90 million more people every year — suggests that we are on the back of a tiger.

We now have to look at the repercussive effects of the industrial revolution on land and land use; on fresh water; on industrial pollution; on the oceans; on the atmosphere, in particular acidification, ozone depletion, and global warming; and on other forms of life.

Can the industrial revolution be extended to the whole planet? If not, what is development for? What should be on the agenda for the Earth Summit?

Almost any forward look compels the conclusion that we cannot continue as we are, and that was fully recognized by the Brundtland Commission. We face not the end of Nature (the foolish title of a recent book), but a change in Nature, in many ways an acceleration of the processes of life. In looking to the future we must reckon with:

- our alarming degree of ignorance. We simply do not know enough about how the world works. Certainly one of the aims of this conference is to help find out. Much current science is about detail and the short term. Few even try to encompass the scene as a whole; they are often regarded with suspicion when they do.

- the character of much change. We tend to see change as something gradual. But critical change is often abrupt. It proceeds by steps or thresholds instead of slopes. We bounce rather than progress from one apparently stable state to another, and the bouncing (and necessary readjustment) can be extremely painful for those around at the time.

- the prospect of surprises. Most people feel that something will always happen to stave off disaster; but disasters have happened in the past, and will happen again. The discovery of ozone holes was entirely unexpected, a point made by the President's Science Adviser.

But ignorance and uncertainty are no excuse for not making the best judgements we can and taking action where necessary or possible. There are certain obvious catalysts which affect both the quality of human life, and life itself. The most obvious is the impact of population increase.

It would be pointless to try and work out all the consequences. But two stand out of particular importance. First we should expect a great increase in human displacements. In 1978 there were something like 5 million refugees in the world on a narrow political definition. I believe there are now more than 17 million. If we add in some 10 million environmental refugees or economic migrants, it means that at present there are around 25 million refugees worldwide. With disruption of current patterns of life, that number would increase dramatically. Sea-level rise would have particularly big effects. It is not fanciful to estimate that with world population rising to 8 billion or more, the refugee rate could rise proportionately with alarming consequences for the integrity of human society as a whole.

There would also be more direct risks to human health with changes in existing patterns of disease. Temperature and moisture are both critical to the ability of viruses, bacteria and insects to multiply. Thus we could see the spread of diseases we had thought under control or far away. Nor should we forget that with loss of biodiversity, it will be less easy to tap the natural world for the constituents of drugs to cope with changes in bacterial and viral populations.

After this visit to the Chamber of Horrors, you may well wonder why the governments and peoples of the world are not concerting action to meet an unprecedented range of problems. The answer is that they are trying to do so. The agenda for the Earth Summit includes an Earth Charter; conventions, agreements or protocols on climate change, biodiversity, and forestry; and an Agenda 21 to set out a list of actions necessary in the next century.

But of course these raise wide issues reading into the better management of our planet. In order to give discussion some precision, I mention four issues which came up at a meeting of the Aspen Institute which I chaired last July:

- The question of institutions and instruments:

    i)   technical ones already exist: for example, the United Nations Environment Program, the World Bank, the World Meteorological Organization and the World Climate Program;

    ii) then there are the central organs of the United Nations: the General Assembly, the Security Council, and the International Court of Justice;

    iii) but we may have to consider something new: for example an Ecological Security Council; a Commission for Sustainable Development; or a series of GATT-type mechanisms arising from the specific conventions.

- The question of finance: we need to make better use of existing resources, but also look at something new (e.g., Global Environmental Facility, and Multilateral Fund under the Montreal Protocol);

- The question of legal issues with the possibility of major development of international environmental law;

- The question of technology and how best to transfer and diffuse it.

Unfortunately, the preparations for this Conference are not going well. As I said at the beginning, countries do see these problems very differently. In a way, "environment" has been captioned by the industrial countries, and "development" by the non-industrial ones, and each is talking a different language. Polarization has followed, and the sad old rhetoric of the 1970s has revived. Why? In my view the main reasons are:

- The industrial countries tend to think of the problem as one which involves them much less than the rest of the world. In fact,

    i)   they are directly or indirectly (and however unwittingly) responsible for the mess (for example, 70% of carbon emissions, 23% from the United States alone);

    ii) their consumption patterns are a root cause, and so far they are unchanged. Such patterns are of course an example that other countries wish to follow;

    iii) by not giving an example of restraint but continuing to preach it to others, they lack credibility;

    iv) they are as guilty as anyone in continuing to use economic instruments and methods of thought that fail to take account of the environmental dimension.

- For their part, the non-industrial and other countries believe — in my view wrongly — that they have the industrial countries over a barrel; somehow they see a bargain in which they trade some measure of environmental restraint against massive quantities of aid, debt forgiveness, new terms of trade, etc. They too are guilty, although more forgivably, for using outdated economic analysis and policies. Many still see the concern in industrial countries about the environment as a trap to slow their development, and keep them in subservient poverty.

I am afraid that both sets of countries nourish illusions. Environment concerns them all, and so does development. They have yet to understand the gravity of the problem. The interests they share are vastly greater than those which divide them. Most important, environmental change will affect all, but be more damaging to poor countries than to rich ones. They are supremely vulnerable.

By contrast, most of the industrial countries should be able to manage over time. But they could scarcely prosper in an impoverished world when they would be a shrinking proportion of the population and where they would be vulnerable to invasion, infiltration and the growing pollution of others. This was an essential point in the NAS Panel on the Policy Implications of Greenhouse Warming to which Dr. Bromley referred. In fact, wealth and poverty cannot sit together for long. Yet the gap between rich and poor is ever widening. According to Swaminathan:

- in 1880, the ratio of real per capita income between Europe on one hand and China or India on the other was 2:1
- by 1965 the ratio was 40:1
- in 1991 it was 70:1.

Thus all need a new international regime. But those who need it most are the poor. They should be the leaders rather than the laggards.

In relation to the size and scope of the issues, we have hardly started to cope with them. We need not only to behave differently but to think differently:

- We need to recast our vocabulary. Words are not only a means of expression but also the building blocks of thought. The instruments of economic analysis are blunt and rusty. Such words as "growth," "development," "cost benefit analysis," even "gross national product," are used in such a misleading way that they are more than ripe for redefinition;
- We need to realize that conventional wisdom is sometimes a contradiction in terms.
- We need to change the culture. Many have lamented the division between the cultures of science and the arts. They are right to do so. But neither is now in charge. Our real bosses are the business managers. Their calculations are strictly short term.
- We need to recast parts of our educational systems to promote better understanding of the environment. In some countries this has begun already, and in others it was always there. In general, I have found the young more in tune with nature than their teachers.
- We need a value system which enshrines the principle of sustainability over generations. Here the Brundtland report has made a beginning. But we have far to go. The proposed Earth Charter for the Summit next year is the first test.

By comparison, behaving differently is almost easy. It follows naturally from thinking differently. Little is possible without a vigorous public opinion putting pressure on local authorities and governments. But the governments themselves are best placed to exercise leadership. There are five main areas for action: over people, over generation and use of energy, over use of land, over industry, and over biodiversity or other forms of life.

For governments a measure of internal reorganization is necessary: for example, there should be tight coordination at the center to ensure integration of policies; environmental audit within Departments, and environmental accounting in annual budgets; environmental costing and pricing with full social cost taken into account; and use of fiscal incentives and disincentives.

All this would amount to a refashioning of our society. But the prospect need not stun us into inaction. I suggest five principles on which governments should act, singly and together.

First, they should now do what makes sense for reasons other than any one environmental factor: the so-called "no regrets" policies.

Next, they should take out insurance policies against disaster, and pay the necessary premiums in terms of precautionary investment.

They should re-target and give more financial support to relevant scientific research and coordinate the results.

They should work out an international strategy which recognizes the realities, sets the framework for collective action, takes good account of equity, and above all is founded on national as well as international interest.

Last, they should deal with environmental issues together. Isolated measures to cope with one of them can sometimes make others worse.

I have left the two most difficult points to grasp to the end:

- We forget that human societies are inherently fragile: not just that of the Soviet Union but in Africa and elsewhere. We do not want to suffer the fate of preceding civilizations.

- We are hooked on the idea of progress, the thought that whatever the setbacks or deviations, human beings advance upwards and onwards. Hence our engrained feeling that problems can always be solved, resources should be exploited to the full, dynamism is better than stability. Certainly this has been a most powerful energizer. Yet in the longer term this philosophy requires at least modification. Sustainability — or sustainable development — carries the implication of balance with nature over an indefinite number of generations. Thus animal species, including our own, and the ecosystems of which they are part, must eventually achieve a stability in which population and resources are in broad equilibrium. Malthus may have got his methodology wrong but his insights were right. Steady-state societies have a tendency to atrophy. Rather we need a steadier state in which our species can accommodate change instead of being endangered, damaged or destroyed by it. Only thus can we come to terms with the realities of our planet.

Appendices

CHEMRAWN VII
ORGANIZING COMMITTEE

General Chair

Professor Robert E. Sievers
Director, CIRES
Department of Chemistry Biochemistry
University of Colorado
Boulder, CO 80309-0216

Members

Dr. Robert M. Barkley
Senior Research Associate
Department of Chemistry Biochemistry
and CIRES
University of Colorado
Boulder, CO 80309-0216

Professor John W. Birks
Future Actions Committee Chair
Department of Chemistry Biochemistry
and CIRES
University of Colorado
Boulder, CO 80309-0216

Dr. Nyle Brady
United Nations Development Program
1889 F St. N.W.
Washington, DC 20006

Dr. Jack G. Calvert
Program Chair
Atmospheric Chemistry Division
National Center for Atmospheric Research
P.O. Box 3000
Boulder, CO 80307-3000

Dr. Eldon E. Ferguson
Director, Climate Monitoring and Diagnostics Laboratory-R/E/CG
U. S. Department of Commerce
National Oceanic & Atmospheric Administration
Environmental Research Laboratories
325 Broadway
Boulder, CO 80303

Mr. Thomas E. Graedel
Poster Committee Chair
Distinguished Member of Technical Staff
AT&T Bell Laboratories
Room 1D-349
600 Mountain Avenue
Murray Hill, NJ 07974

Dr. Rudolph Pariser
Vice-Chair and CHEMRAWN Committee Representative
Science Director, Du Pont (Retired)
851 Old Public Road
Hockessin, DE 19707

Professor Cyril Ponnamperuma
Third World Chair
Department of Chemistry
University of Maryland
College Park, MD 20742

Dr. Bryant Rossiter
Eastman Kodak (retired)
25662 Dillon Road
Laguna Hills, CA 92653

Dr. William G. Schneider
National Research Council of Canada
100 Sussex Drive
Ottawa, Ontario
Canada K1A OR6

Dr. William E. Wilson
Workshop Chair
U.S. Environmental Protection Agency
Mail Drop 75
Atmospheric Resarch
Research Triangle Park, NC 27711

Staff

Dr. John M. Malin
Awards
International Activities
American Chemical Society
1155 Sixteenth St., N.W.
Washington, D.C. 20036

Ms. Julie McKie
CIRES
University of Colorado
Boulder, CO 80309-0216

Ms. Christine P. Pruitt
Department Head
Mtgs. and Div. Act. Depart.
American Chemical Society
1155 Sixteenth St., NW
Washington, DC 20036

Ms. Diane Ruddy
Mtgs. and Div. Act. Depart.
American Chemical Society
1155 Sixteenth St., NW
Washington, DC 20036

CHEMRAWN VII
PROGRAM COMMITTEE

Dr. Jack G. Calvert, *Chair (National Center for Atmospheric Research, USA)*
Dr. Hajime Akimoto *(National Institute of Environmental Studies, Tsukuba, Japan)*
Dr. Nicolai M. Bazhin *(Institute of Chemical Kinetics and Combustion, Novosibirsk, Russia)*
Dr. Richard H. Brown *(Occupational Hygiene Lab., London, England)*
Dr. Richard A. Cox *(Atomic Energy Research Establishment, Harwell, England)*
Dr. Karl H. Eickel *(Verein Deutscher Ingenieure Kommission Reinhaltung der Luft, Düsseldorf, Germany)*
Dr. Eldon Ferguson *(NOAA, Boulder, Colorado, USA)*
Dr. Georgii Sergeevich Golitsyn *(Institute of Atmospheric Physics, Moscow, Russia)*
Dr. Thomas Graedel *(AT&T Bell Labs, Murray Hill, New Jersey, USA)*
Dr. H.J. Grosse *(Zentralinstitut für Isotopen und Strahlenforschung, Leipzig, Germany)*
Dr. Øystein Hov *(Geophysics Institute, University of Bergen, Bergen, Norway)*
Dr. J. Alistair Kerr *(Swiss Federal Institute, EAWAG Technical University, Zürich, Switzerland)*
Dr. Leo Klasinc *(Rudjer Bosković Institute, Zagreb University, Zagreb, Yugoslavia)*
Dr. J.-E.O. Lindqvist *(Göteborg University, Göteborg, Sweden)*
Dr. Mack McFarland *(E.I. du Pont de Nemours & Co., Wilmington, Delaware, USA)*
Dr. U. Ozer *(Uluday Universitesi, Bursa, Turkey)*
Prof. Stuart A. Penkett *(University of East Anglia, Norwich, England)*
Prof. Leon F. Phillips *(University of Canterbury, Christ Church, New Zealand)*
Dr. Eugenio Sanhueza *(Cent. Ing. Venez. Invest. Cient., Caracas, Venezuela)*
Prof. Harold I. Schiff *(York University, Ontario, Canada)*
Dr. Onkar N. Singh *(Applied Physics Institute of Technology, Varanasi, India)*
Dr. Sjaak Slanina *(Netherlands Energy Research Foundation ECN, Petten, Netherlands)*
Prof. Xiaoyan Tang *(Beijing University, Beijing, China)*
Dr. Peter Warneck *(Max Planck Instute für Chemie, Mainz, Germany)*
Dr. William E. Wilson *(Atmospheric Research and Exposure Assessment Lab., Environmental Protection Agency, Research Triangle Park, North Carolina, USA)*
Dr. V.E. Zuev *(Institute for Atmospheric Optics, Tomsk, Russia)*

CHEMRAWN COMMITTEE

Officers

Sir J.M. Thomas, *Chair (Royal Institution of Great Britain, United Kingdom)*
Dr. A. Hayes, *Vice Chair (Imperial Chemical Industries PLC, United Kingdom)*
Prof. R.H. Marchessault, *Secretary (McGill University, Canada)*

Members

Dr. J.B. Donnet *(CRPCSS, France)*
Dr. E.J. Grzywa *(Industrial Chemistry Research Institute, Poland)*
Dr. J. Klein *(GSF - Forschungszentrum für Umwelt und Gesundheit, Germany)*
Dr. R. Pariser *(E.I. du Pont de Nemours & Co. (Retired), USA)*
Dr. S. Varadarajan *(Consultancy Development Centre, India)*

Associate Members

Dr. D.A. Bekoe *(International Development Research Centre, Kenya)*
Dr. F.A. Kuznetsov *(Institute of Inorganic Chemistry, USSR)*
Dr. M. Morita *(National Institute for Environmental Studies, Japan)*
Dr. P. Moyna *(Universidad de la Republica, Uruguay)*
Prof. C. Ponnamperuma *(University of Maryland, USA)*

CHEMRAWN VII
CONTRIBUTORS

Major Contributors

Dow Chemical, USA
E.I. du Pont de Nemours & Company
Monsanto
National Oceanic and Atmospheric Administration
National Science Foundation
Toyota Motor Corporation
United Nations Development Programme
U.S. Environmental Protection Agency

Contributors

Agency for International Development
Allied Signal Inc.
American Cyanamid Company
ARCO Chemical Company
BASF Corporation
BF Goodrich Company
BP America, Inc.
Eastman Kodak Company
Ethyl Corporation
Fiat, USA, Inc.
Hercules Incorporated
Hoechst Celanese Corporation
ICI Americas, Inc.
Mobay Corporation
Nissan Motor Company, Ltd.
Olin Corporation
PPG Industries, Inc.
Sony Corporation
Texaco, Inc.
Union Carbide Corporation
United Technologies

INDEX